# Forest Science - Advances towards Sustainable Development and Climate Resilience

*Edited by Gopal Shukla, Biplov Chandra Sarkar, Zishan Ahmad Wani, Jahangir A. Bhat and Sumit Chakravarty*

Published in London, United Kingdom

Forest Science - Advances towards Sustainable Development and Climate Resilience
http://dx.doi.org/10.5772/intechopen.1006013
Edited by Gopal Shukla, Biplov Chandra Sarkar, Zishan Ahmad Wani, Jahangir A. Bhat and Sumit Chakravarty

Contributors
Aakanksha Pandey, Abel Ramoelo, Abha Manohar Kandileri, Abubakr A.A. Osman, Alma Angelina Hernández Rodríguez, Archana Mishra, Asep Denih, Bilal A. Bhat, Chaitanya Kolekar, Emad H.E. Yasin, Emily García Montiel, Eusebio Montiel Antuna, Fath Alrhman A.A. Younis, Gopal Shukla, Guiwu Zou, Hanjian Hu, Iyaz Quyoom, Jesús Alejandro Soto Cervantes, Kang Xu, Magno Gonçalves-Araújo, Mahlomola Ernest Daemane, Pablito Marcelo López Serrano, Ravindra Medhe, Samuel Adelabu, Smaranika Mohanta, Sumit Chakravarty, Wasim Sajad Malik



First published in London, United Kingdom, 2025 by IntechOpen
IntechOpen is the global imprint of INTECHOPEN LIMITED, registered in England and Wales, registration number: 11086078, 167-169 Great Portland Street, London, W1W 5PF, United Kingdom

For EU product safety concerns: IN TECH d.o.o., Prolaz Marije Krucifikse Kozulić 3, 51000 Rijeka, Croatia, info@intechopen.com or visit our website at intechopen.com.

British Library Cataloguing-in-Publication Data
A catalogue record for this book is available from the British Library

Forest Science - Advances towards Sustainable Development and Climate Resilience
Edited by Gopal Shukla, Biplov Chandra Sarkar, Zishan Ahmad Wani, Jahangir A. Bhat and Sumit Chakravarty
p. cm.
Print ISBN 978-1-83634-713-2
Online ISBN 978-1-83634-712-5
eBook (PDF) ISBN 978-1-83634-714-9

# Meet the editors

Dr. Gopal Shukla is Professor and Head of the Department of Forestry, North-Eastern Hill University (NEHU), Tura Campus, Tura-794002, Meghalaya, India. With over two decades of academic and research experience, he has made significant contributions in agroforestry, biodiversity conservation, sustainable forest management, and the socio-ecological dynamics of forest-dependent communities, particularly in North-East India. His research spans plant diversity assessment, utilization patterns of non-timber forest products (NTFPs), biomass and carbon sequestration, soil health in agroforestry systems, and community-based adaptation to climate change. He has published more than 100 research papers, 30 book chapters, and 8 books, and contributed extensively to policy-oriented technical reports. Dr. Shukla has successfully led and collaborated on numerous government- and externally-funded projects, integrating ecological field studies with participatory rural appraisal and indigenous knowledge documentation. He has supervised four doctoral and thirteen master's students, fostering the next generation of forestry professionals. Actively engaged in the global scientific community, Dr. Shukla serves on the editorial boards of leading journals, including *Advances in Bamboo Science*, and has been a reviewer and guest editor for several special issues. His work exemplifies the integration of scientific innovation with grassroots conservation to promote sustainable natural resource management in ecologically fragile regions.

Dr. Biplov Chandra Sarkar is an Assistant Professor at the Faculty of Agricultural Sciences, GLA University, Mathura, Uttar Pradesh. He earned his Ph.D. in Forestry (2020) and M. Sc. in Forestry (2016) from Uttar Banga Krishi Viswavidyalaya, and his B. Sc. in Forestry (2014) from SHUATS, Uttar Pradesh. He is ASRB NET qualified in Agroforestry. Dr. Sarkar has published 20 peer-reviewed articles and six book chapters. He received the "Best Ph.D. Thesis Award 2022" from the Society for Conservation of Natural Resources, Kolkata. His professional experience includes serving as a Guest Lecturer at NEHU, Tura (2021–2023) and a Technical Officer (Forestry) at MoEF&CC, Kolkata (2020–2021). His work reflects a strong commitment to forestry education, research, and policy.

Dr. Zishan Ahmad Wani is a plant ecologist currently working as a Post-Doctoral Fellow at SRM University–AP, India. He holds a Ph.D. in Botany from Baba Ghulam Shah Badshah University, Rajouri, with earlier degrees in Botany (M. Sc. from HNB Garhwal Central University) and Education (B. Ed. from University of Kashmir). His research focuses on floristics, plant biodiversity, species distribution modelling, ecosystem services, and ethnobotany, with a strong emphasis on sustainability and climate resilience. Dr. Wani has previously served as a Research Associate at GB Pant National Institute of Himalayan Environment and as a Senior Research Associate at ATREE, Bangalore. He has also taught subjects such as ecology, conservation biology, and environmental sciences at both undergraduate and postgraduate levels. Dr. Wani has authored over 45 research papers in reputed international journals, contributed multiple book chapters with publishers like Springer and Elsevier, and co-authored a book on economically important plants. He is a reviewer for several high-impact journals, including Scientific Reports, Sustainability, and Frontiers in Plant Science. His work has received over 550 citations, with an h-index of 14. He is also a member of the IUCN Commission on Environmental, Economic and Social Policy (CEESP) and the Ecosystem Restoration Alliance–India.

Dr. Jahangeer A. Bhat is currently associated with the Piramal Foundation for Education Leadership at Piramal University, Mumbai, India. Before taking on his current position, he was associated with the College of Horticulture and Forestry at Rani Lakshmi Bai Central Agricultural University, India. Previously, he served as the head of the Department of Forestry at the College of Agriculture, Fisheries and Forestry, Fiji National University, Republic of Fiji Islands. His research work in India and the Fiji Islands spans 12 years and encompasses emerging issues in forestry, including conservation of biodiversity, traditional knowledge of plants, and sustainable management of forest resources, particularly in the light of climate change scenarios. He has worked as a counselor, mentor, and coordinator for forestry academic programs. He has been instrumental in developing higher education and technical and vocational education and training streams of forestry and allied programs. He worked closely with the Fiji Higher Education Commission and forestry stakeholders for accreditation.

Dr. Sumit Chakravarty is Professor and Head of the Department of Forestry, Uttar Banga Krishi Viswavidyalaya, Pundibari, Cooch Behar, West Bengal, India. He has more than 20 years of experience in forestry training, research, and development. He has researched several aspects of forestry, agroforestry, medicinal plants, and climate change. He has published 120 research papers, 48 book chapters, and 8 books. He is involved in fifteen research projects on forest ecology, medicinal plants, and agroforestry. He has trained over fifty master's and doctoral students in forestry. Dr. Chakravarty has attended more than fifty national and international conferences as a presenter and invited speaker.

# Contents

# Preface

Forests play a vital role in sustaining life on Earth. They support rich biodiversity, regulate climate, and provide essential ecosystem services like clean air, water protection, and soil conservation. Forests also serve as a lifeline for many communities, especially in rural and tribal areas, offering food, medicine, fuelwood, and income. Beyond their ecological and social importance, forests play a crucial role in combating climate change by absorbing carbon dioxide, storing carbon, and serving as natural barriers against extreme weather events like floods, droughts, and storms. As the world faces growing challenges like rising global temperatures, biodiversity loss, and increasing pressure on natural resources, the role of forests has become more crucial than ever.

This book, Forest Science – Advances towards Sustainable Development and Climate Resilience, highlights the growing importance of forests and agroforestry in addressing some of today's most pressing issues, including climate action, sustainable development, habitat conservation, food and nutritional security, and the integration of advanced technologies for forest monitoring and management. It brings together knowledge from different fields to explore how forests can help solve some of the most pressing global challenges. The chapters cover a wide range of topics, including scientific advances, climate solutions, sustainable land use, and the social and economic aspects of forest management. Some chapters focus on how forests act as carbon sinks and what strategies can help increase their ability to absorb and store carbon. The book also includes insights into forest certification systems. There are studies on forest health, biodiversity, the impact of climate on forest ecosystems, and how wild edible plants can help improve nutrition in rural areas. Others highlight the use of remote sensing and Internet of Things (IoT) technologies to monitor forest conditions, detect environmental changes, and support better decision-making.

Our goal with this book is to provide valuable insights for researchers, students, policymakers and anyone interested in forests and sustainability. We believe that by understanding and managing our forests better, we can make meaningful progress toward a healthier planet and a better future for all.

We sincerely thank all the contributors for their hard work and dedication. We hope this book inspires readers to appreciate the value of forests and take positive steps toward sustainability and resilience.

**Gopal Shukla**
North Eastern Hill University,
Tura Campus,
Tura, Meghalaya, India

**Biplov Chandra Sarkar**
GLA University,
Uttar Pradesh, India

**Zishan Ahmad Wani**
Department of Environmental Science and Engineering,
SRM University,
Andhra Pradesh, India

**Jahangeer A. Bhat**
Piramal Foundation for Education Leadership,
Piramal University,
Mumbai, India

**Sumit Chakravarty**
Uttar Banga Krishi Viswavidyalaya,
Pundibari, Cooch Behar (West Bengal), India

Section 1

# Climate Change Mitigation and Carbon Sequestration

Chapter 1

# Forest Carbon Sequestration Functions and Mitigation Strategies for Global Climate Change

*Kang Xu, Guiwu Zou and Hanjian Hu*

## Abstract

This chapter aims to explore the critical role of forests in mitigating global climate change by enhancing carbon sequestration. We synthesized the research of global forests carbon sink, showcasing their significant carbon sink potential. Drawing on previous studies, we introduced the contributions of various forest types and management practices to bolster forest carbon sequestration. Case studies of successful nature-based solutions (NbS) projects are highlighted to illustrate effective strategies for increasing forests carbon sinks. Additionally, we outlined methodologies for quantifying forest carbon sequestration, encompassing field-based approaches, remote sensing technologies, and advanced modeling techniques. Sustainable forest management strategies were evaluated, with a focus on afforestation, reforestation, conservation, and efficient utilization of forest resources, including wood products and renewable energy integration. Furthermore, we discussed existing policy frameworks, with a particular emphasis on the REDD+ initiative, and provide recommendations to enhance their effectiveness in promoting forest carbon sequestration. In summary, this chapter offers a holistic perspective on how forests can be leveraged as a vital tool in combating climate change.

**Keywords:** carbon sequestration, carbon neutrality, climate change mitigation, forest management, greenhouse gas

## 1. Introduction

Climate change, driven by the relentless increase in atmospheric carbon dioxide ($CO_2$) and other greenhouse gases (GHGs), poses a significant threat to global ecosystems and human societies. Plant photosynthesis, a vital process in the carbon cycle, serves as the sole pathway for $CO_2$ to transition from the atmospheric reservoir into biological organisms. The process, activities, and mechanisms that remove $CO_2$ from the atmosphere and store it for a period are commonly referred to as carbon sinks. Photosynthesis therefore stands as the largest natural carbon sink channel, playing a crucial role in mitigating the impact of climate change. It is well-established that the escalating GHG emissions, leading to an expansion of the atmospheric carbon pool, are the direct driver of global climate change. To achieve the ambitious goal of limiting global warming to 2°C or below, or even 1.5°C, it is insufficient to rely solely on reducing emissions from fossil fuels and land-use changes. We must also broaden our approach to enhance carbon sinks, thereby paving the way for ultimate carbon neutrality.

Global terrestrial carbon sinks, estimated at approximately 2.3 ± 1.0 Gt C $yr.^{-1}$, could offset 23% of emissions from fossil fuel, thus slowing the pace of global climate warming [1]. Forests, the land type with the largest biomass, play an indispensable role in its function of carbon absorption against climate change. When accounting for the carbon sink loss due to deforestation in tropical rainforests, global forest sink has the potential to nearly offset half of fossil fuel emissions [2].

Therefore, it is crucial to understand, utilize, and tap into the full potential of forest carbon sequestration. Not only does it involve protecting existing forests, but it also involves promoting reforestation, afforestation, and sustainable forest management practices, as well as various ways of utilizing resources. In this chapter, we aim to: (1) highlight the importance of forest carbon sequestration for carbon neutrality and climate change mitigation, (2) review global and regional forest carbon sink potential and influencing factors, (3) detail various measurement methods for quantifying forest carbon sequestration, and (4) discuss sustainable forest utilization strategies, including carbon resource utilization through wooden products, renewable energy integration, and supporting policy frameworks. Integrating research from ecology, forestry, remote sensing, and policy analysis, this chapter focuses on the potential of forests as carbon sinks, methods for quantifying this potential, and strategies for enhancing and sustainably utilizing forest carbon sequestration to contribute to global climate stability and sustainable development.

## 2. Global forest carbon sink potential

### 2.1 Review of recent research papers on forest carbon sequestration potential

#### *2.1.1 Carbon sink of global forests and their potential*

Forests have a crucial role in carbon sequestration through the process photosynthesis, which usually leads to their classification as carbon sinks. Some studies take the opposite view, arguing that forests are carbon sources at both global and regional scales. Baccini et al. [3] quantified the net annual changes in the aboveground carbon density of tropical woody live vegetation using the moderate resolution imaging spectroradiometer (MODIS) pantropical satellite data and argued that the world's tropical forests are a net carbon source of 425 Tg C $yr.^{-1}$ (1 Tg = $10^{12}$ g) from 2003 to 2014. Similarly, Dixon et al. [4] demonstrated that the net carbon emission of global forests is 0.9 ± 0.4 Pg C $yr.^{-1}$ (1 Pg = $10^{15}$ g). However, mounting evidence supports the view that forests are carbon sinks at a global scale, with carbon source status limited to specific regions or periods [2, 5, 6]. For example, the young forests in southern Sweden switch from carbon source to carbon sink after eight years of reforestation [7]. Intriguingly, previous studies that argued forests as carbon sources have proven that forests are carbon sinks. Baccini et al. [3] noted that the net carbon release from tropical forests comprised losses of 861.7 ± 80.2 Tg C $yr.^{-1}$ and gains of 436.5 ± 31.0 Tg C $yr.^{-1}$, with the former resulting from forest growth and the latter from deforestation and decreased carbon density due to degradation or disturbance. The 0.9 Pg C $yr.^{-1}$ net flux to the atmosphere of the global forest is the result of 1.6 Pg C $yr.^{-1}$ emission from deforestation in low latitudes and 0.7 Pg C $yr.^{-1}$ sequestration from forest area expansion and growth [4]. A close inspection reveals that the reason why forests are considered a carbon source is because of deforestation.

As the main component of terrestrial ecosystem, forests account for over 80% of the terrestrial ecosystem's carbon sinks and sequester approximately 25% of anthropogenic carbon emissions [8, 9]. Over time, the estimated magnitude of global forest carbon sinks has increased. Based on the ground-based measurements, Pan et al. [2] demonstrated that the global forest is an enduring and steady carbon sink, segregating 3.6 ± 0.4 Pg C $yr.^{-1}$ and 3.5 ± 0.4 Pg C $yr.^{-1}$ in the 1990s ~ 2000s and the 2010s, respectively, and neutralizing almost half of fossil-fuel emissions (7.8 ± 0.4 Pg C $yr.^{-1}$ in 1990 ~ 2019). Meanwhile, using the aboveground biomass, Alaniz et al. [10] indicated that the global forest is a carbon sink segregating 0.06 ± 0.58 Pg C $yr.^{-1}$ from 2000 to 2019. Integrating ground and Earth observation data, Harris et al. [11] estimated that global forests were a net carbon sink of 2.1 ± 13.4 Pg C $yr.^{-1}$ during 2001–2019.

In addition to ground measurement, many researchers have estimated forest carbon sinks globally or regionally using processed-based models. According to "Global Carbon Budget 2024," global forests sequestered 1.3 Pg C $yr.^{-1}$ during 2013–2022 based on dynamic global vegetation models (DGVMs). Another study, also based on DGVMs, found a carbon sink of 2.15 Pg C $yr.^{-1}$ [12]. By upscaling eddy covariance data with a random forest, Zeng et al. [13] demonstrated the carbon sink was 0.4 Pg C $yr.^{-1}$ from 1999 to 2019. Using $CO_2$ observations and inverse models, Schimel et al. [14] reported that the carbon sink was 1 ~ 2 Pg C $yr.^{-1}$ during the 2000s. In addition, there is a large number of studies that apply atmospheric inversions and flux measurements to estimate forest carbon sinks at the regional scale [15]. Collectively, these studies indicate that global forests function as carbon sinks, albeit they may locally act as carbon sources.

Despite their current status as the primary carbon sink in terrestrial ecosystems, forests possess significant potential for enhanced carbon sequestration through management practices. The global forest area has been declining over the past few decades (**Figure 1**), whereas the carbon sink capacity of forests remains enduring [2]. Through improved management, the existing global forest could offer a carbon sink of 206 Pg C, with the majority (71%) concentrated in tropical forests [17]. Following the 2015 Paris Agreement, parties to the UNFCCC agreed to hold global warming to "well below" 2°C and pursue efforts to limit it to 1.5°C by reducing global $CO_2$ emissions as soon as possible and reaching net-zero by mid-century. Many countries have already taken action. For example, China proposed achieving a carbon peak in 2030 and carbon neutrality in 2060. Under a tree-planting scenario aligned with China's ecosystem management policy, China's forests can sequester 5.9 ± 0.5 Pg C [18]. Thus, the global forest retains substantial potential for carbon sequestration.

#### *2.1.2 Role of different forest types and management practices in enhancing carbon sequestration*

*Tropical forests* are the most efficient carbon sinks due to their rapid growth rates and high biodiversity. Despite covering only 30% of the global tree cover, these forests harbour 50% of the world's carbon stored in trees (WRI, Research. (n.d.). Forest Carbon Stocks. https://research.wri.org/gfr/biodiversity-ecological-services-indicators/forest-carbon-stocks#how-much-carbon-is-stored-in-the-world-s-forests).

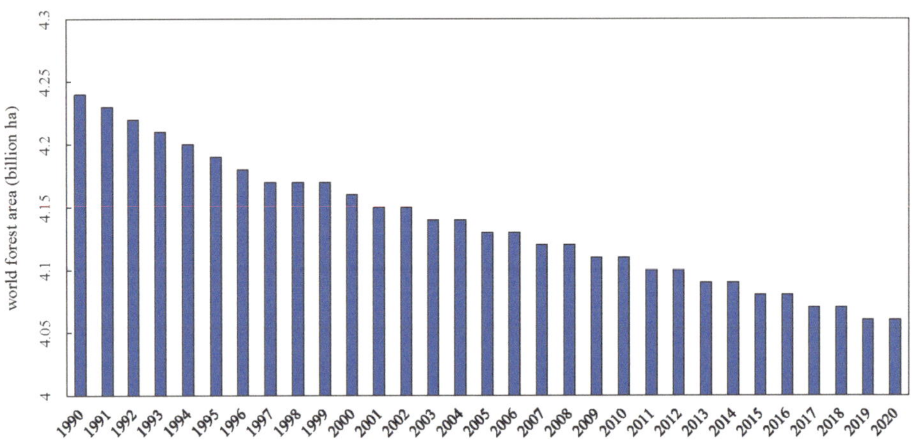


**Figure 1.**
*The global forest area over 1990–2020 (cited from [16]).*

The diverse species composition allows for complex interactions that enhance growth and resilience, which in turn boosts carbon storage [19]. Meanwhile, the intense deforestation offsets much of their carbon sequestration. *Temperate forests* also play a significant role in carbon sequestration, although their rates are generally lower than those of tropical forests. In temperate forests, soil organic carbon (SOC) accounts for 60% of the ecosystem carbon stock [20]. The rapid expansion of forest area in this region, particularly through national-scale afforestation and reforestation programs in China, has been instrumental in enhancing carbon sinks [2]. *Boreal forests* have slower growth rates but segregate vast amounts of carbon in soil (84%) and dead biomass due to the decomposition rate at low temperatures [21]. However, the increasing incidence of wildfires, driven by climate warming and drying, poses a great threat to the carbon sink capacity of boreal forests [22]. From 1948 to 2014, wildfires resulted in 789 Tg C emission in North American boreal forests alone [23]. Meanwhile, the rising outbreaks of insects due to climate warming exacerbate carbon emissions from living biomass. Reducing wildfires and insect outbreaks is crucial for enhancing carbon sinks in boreal forests.

Enhancing forest carbon sequestration involves two primary strategies: increasing forest area and elevating the carbon density of forests per unit area through management practices. Over the past two decades, China has carried out extensive afforestation activities. Afforestation efforts in southern China during the last two decades sequestered approximately 2.34 Pg C between 2002 and 2017, with a potential for an additional 5.32 Pg C sequestration [24]. Maximizing the carbon sink potential per unit area of forests is equally important. Selecting forest species that have the maximum fitness to the local environment and maximal carbon stock enabled almost a doubling in forest carbon sink potential [25]. Regarding tree diversity and individual density, the diversity of species is ideal for recapturing carbon in tropical and equatorial forests, whereas the abundance of trees is more important to maximize carbon storage in cold and dry regions [19].

Forest management plays a pivotal role in encouraging trees to sequester more carbon, by accelerating tree growth and soil carbon accumulation. The soil fertility is often the limitation for forest net primary production (NPP) [26]. Fertilization is a widely used management practice in intensive forest management. Forests in marginal sites, due to their low soil quality, exhibit significantly lower carbon accumulation compared to forests in non-marginal lands [27]. It is important to acknowledge that while fertilization may temporarily promote respiration, it ultimately enhances the forest's carbon sink. The first fertilization in young Norway spruce forest initially increases forest floor respiration but the enhancement of photosynthesis exceeds the floor respiration [28].

To improve tree growth and regeneration, thinning—the selective removal of trees using single-tree or group-selection harvesting methods—can be employed. Although the thinning reduces the carbon storage in vegetation and accelerates soil $CO_2$ emission due to the disturbance, recent research has demonstrated that moderate thinning increases the total carbon density of forest by accelerating the tree growth and by increasing the recalcitrant fraction of soil carbon [29, 30]. Using wood products instead of other products that require the use of fossil fuels (e.g., plastics) is also a by-product that enhances forest carbon sink (details in Chapter 4.2). Compared to letting trees die and decompose, they can be harvested and processed into products like garnish and building materials which can be around for decades or even centuries. The Liberty Bell is a quintessential example; its wooden yoke, made from American elm harvested in the 1770s, still stores carbon today.

Promoting species and structural diversity in forests is another effective and commonly used management strategy. The positive relationships between species diversity and forest productivity have been widely evidenced [31, 32]. A study based on Canada's National Forest Inventory (NFI) database revealed that increased species evenness increased soil carbon storage by 30% [33]. Recently, many researchers have emphasized the importance of structural diversity over species diversity in enhancing forest productivity [34–36]. Therefore, enhancing structure diversity is a better way to improve forest carbon sink. Simultaneously, the high species diversity also reduces carbon losses by enhancing their resilience (resistance, recovery, and adaptability) to various disturbances such as wildfires, insects, drought, and biological invasion [37].

### 2.2 Successful NbS projects that have increased forest carbon sinks

Climate change is one of the greatest challenges that mankind has ever faced, posing significant threats to the environment and nearly all aspects of human well-being [5]. The solutions will not be immediate, even if organized in an integrated and global manner. Against this backdrop, nature-based solutions (NbS) are emerging as an integrated approach to solving this issue [38]. NbS was defined as "actions to protect, sustainably manage, and restore natural or modified ecosystems that address societal challenges effectively and adaptively, simultaneously providing human well-being and biodiversity benefits" [39]. Girardin et al. [40] claimed that implementing NbS such as forest protection could contribute to a 0.1°C reduction in global warming by mid-century, under a scenario where warming peaks at 1.5°C by that time.

Numerous successful cases illustrate the effectiveness NbS in enhancing forest carbon sinks. The Three-North Shelterbelt Program (TNSP), known as China's "Great Green Wall" (GWC), has been developed for over 40 years and is one of the most successful NbS cases. Since 1978, this program has been deployed by China to fight the encroaching Gobi Desert and help the fortunes of its northern provinces. Acting as a buffer between the Gobi and the rest of the mainland, the GWC is in the middle of swelling to more than 100 billion trees along a 4500-km belt by 2050. It will eventually stretch over one-tenth of the country. The cumulative carbon sequestration of the TNSP was 2.31 Pg C from 1978 to 2017, which corresponded to 5% of China's industrial carbon emissions in the same period [41]. Despite mixed assessments, over 66 billion trees had been planted by 2014, and scientists attribute the TNSP's success to repelling and, in some instances, reversing desert encroachment. By planting native species well-adapted to the local climate, the area has transformed into an oasis, leading to reduced local temperatures, increased soil carbon, and the creation of new forest-related incomes in numerous rural communities. China has witnessed a substantial increase in forest cover and associated carbon stocks, with the carbon sink of the GWC continuing to grow [42].

Another notable NbS case is the Southern Cardamom REDD+ Project in Koh Kong Province, Cambodia. The Cardamom Rainforest Landscape is one of the last unfragmented rainforests remaining in Southeast Asia and is a critical part of the Indo-Burma Biodiversity Hotspot. The government pay farmers 70% above the normal market price for their organic rice and forest stewardship, and hence prevents farmers from clearing new land illegally in the forests. Using global best practices for forest protection and community development, the Southern Cardamom REDD+ project prevents more than 3 Tg C emissions annually (https://ibisrice.co.uk/business-model/).

It is important to acknowledge that NbS may eventually reach a saturation point where the ecosystem attains equilibrium, and sequestration is balanced by emissions. For example, after two decades of afforestation in southern China, the growth rate of carbon sink for these forests has begun to slow, and 71% of its carbon storage potential has been achieved by 2017 [24].

### 2.3 Challenges and opportunities for scaling up forest carbon sequestration efforts globally

Forest carbon sequestration stands as a pivotal strategy in the battle against climate change. While forests offset approximately 25% of anthropogenic GHG emissions, enhancing their carbon sequestration capacity undoubtedly holds promise for mitigating global climate change [9]. Scaling up efforts in forest carbon sequestration presents a myriad of challenges and opportunities. Understanding these factors is essential for enhancing the carbon sinks of forests.

One of the most parlous challenges is deforestation and land use change (LUC), which often switch forest from carbon sinks into long-term carbon sources. Agricultural expansion, logging, infrastructure development, and urbanization are primary drivers of deforestation, significantly reducing forest cover and carbon storage capacity [43]. From 1990 to 2019, the carbon emission from deforestation in tropical forests is 2.24 Pg C $yr.^{-1}$ [2]. Furthermore, deforestation can exacerbate ecosystem stress, increase the incidence of fires, and lead to higher carbon emissions [44]. Although the rate of deforestation is declining (**Figure 2**), we are still far from achieving zero deforestation.

Another shriveled challenge is the immediate and inescapable impact of climate change itself. The intensification of the dry season and pest outbreaks accelerated by warming temperatures cause a great deal of trees to die [45, 46]. Moreover, the drought also triggers forest fires [47], further weakening the carbon sink of forests. Meanwhile, global warming also enhances soil carbon emissions [21].

The third challenge is carbon sink verification and funding constraints. Forest carbon sequestration is a long-term and complex process; hence, accurately measuring and verifying carbon sinks in forests can be complex and resource-intensive, posing challenges for reporting and accountability in carbon offset programs. Meanwhile,

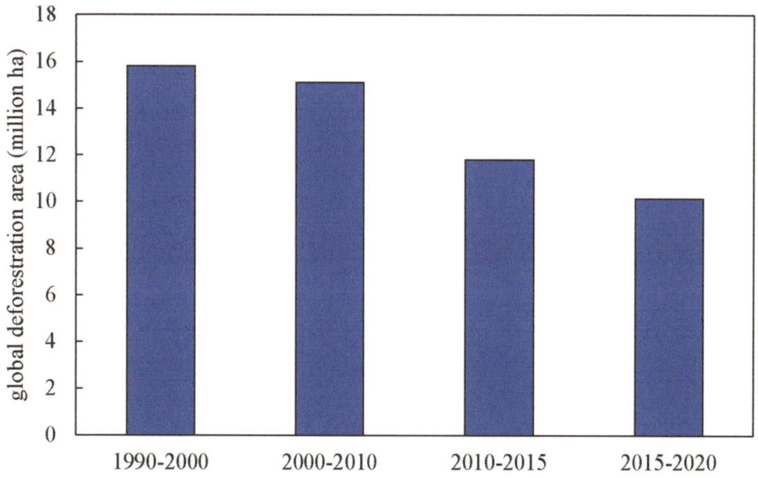


**Figure 2.**
*The global deforestation area over 1990–2020 (cited from [16]).*

substantial investments are needed for large-scale reforestation, afforestation, and sustainable land management. In 2024 alone, the Chinese government have invested 32 billion Yuan in the construction of the Three-Norths Shelter Forest. Hence, limited access to funding can impede project implementation.

The fourth challenge is land tenure and governance issues. In tropical countries, unclear land tenure can hinder reforestation and sustainable management efforts, as communities may lack the incentive to invest in long-term forestry practices. The areas with the highest recovery potential are disproportionately distributed in countries with weak rule of law, and often in countries with large amounts of unrecognized land tenure. At least 67% of the areas with the highest restoration potential in Madagascar must be on untitled land, where tenure is often unclear or disputed [48]. Inconsistent policies and lack of enforcement can undermine the effectiveness of carbon sequestration initiatives, making it difficult to implement sustainable practices [49].

Integrating forest carbon sequestration into national climate action plans, such as Nationally Determined Contributions (NDCs) under the Paris Agreement, can provide essential support and funding for conservation and restoration projects. Several opportunities are provided against these challenges. The first opportunity lies in conserving and enhancing forest land [50]. There is still potential forestation land, for example, the area of prioritized potential forestation land in China is 0.66 million $km^2$ [51], and the area of high potential for restoration in tropical is 27 Mha [48]. If all these lands are afforested, they can provide immense carbon sinks. The second opportunity is the participation in forest carbon markets. Credits generated from forest carbon projects can be sold in either compliance or voluntary carbon markets. This market provides different entities with the option to purchase forest carbon offsets to meet emission reduction targets and allows for lower-risk investment in forest carbon projects with strong incentives for project completion [50]. The third opportunity lies in the wood product market. Timber buildings are low-carbon and energy-saving buildings. Compared with reinforced and concrete buildings, wood structure buildings reduce carbon emissions by 48.9% ~ 94.7% in the production stage of building materials. During the subsequent life cycle, the greenhouse gas emissions of wood-frame buildings are 20% ~ 50% lower than those of steel or concrete structures [52, 53].

Furthermore, leveraging technological advancements, such as remote sensing and satellite imagery, can enhance monitoring and verification of carbon emissions and stocks, ensuring transparency and accountability in carbon markets (more details in Chapter 3.5). This technology even can estimate the carbon content in dead fine wood debris [54]. It can also monitor disturbances in forests such as wildfires, anthropogenic deforestation, and illegal logging in real-time, and therefore, the government can intervene immediately. At the same time, the burgeoning artificial intelligence (AI) technology is likely to bring about a revolution in the field of climate change in the future (details in Chapter 3.2.2 and 3.5). By integrating information with AI, we can achieve efficient and accurate acquisition of multi-scale forest ecosystem parameters [55]. The concept of Climate-Smart Forestry has also been put forward [56].

## 3. Quantifying forest carbon sequestration

Quantifying the carbon sequestration capacity of forests is the basis for evaluating the ecosystem services, functioning, and carbon neutrality of forests. It is of great significance for protecting ecosystem biodiversity and mitigating global climate change, and hence a solution to global climate issues [18]. We introduce the primary methods employed for measuring forest carbon sequestration, including field-based approaches, remote sensing techniques, the eddy covariance method, process-based modeling, and deep learning models (**Figure 3**).

### 3.1 Field-based methods: Use of field data, allometric equations, and growth models

Field-based methods represent the earliest approaches used for estimating forest carbon sequestration, relying on biomass data. Therefore, accurate measurement

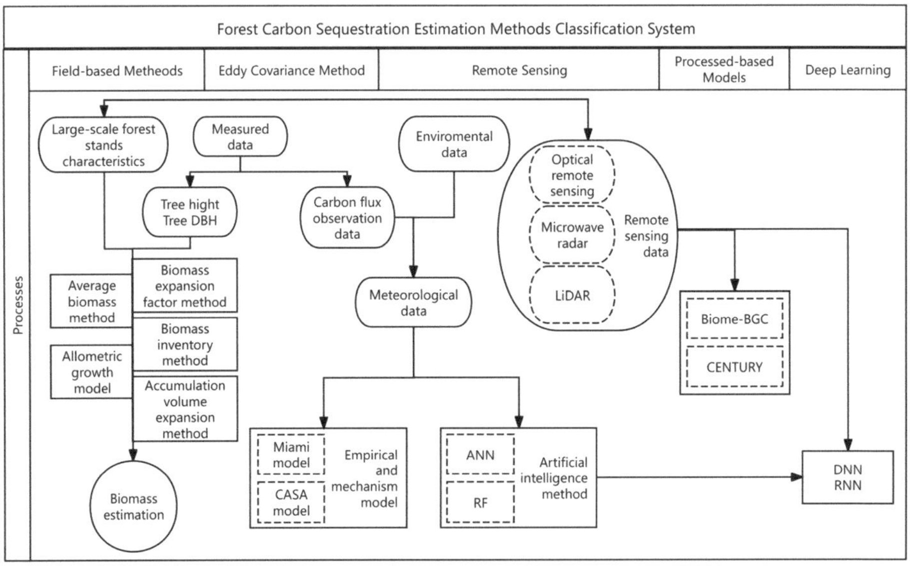


**Figure 3.**
*Forest carbon sequestration estimation methods classification system.*

of biomass has become the key point of these methods. Initially, average biomass method obtain biomass data by direct-harvesting. Datasets consisting the relationship between inventory data and biomass were established with the gradual advancement of parameters for biomass measurement, and biomass expansion factor method, biomass inventory method, and accumulation volume expansion method were proposed based on these datasets.

#### *3.1.1 Average biomass method*

The average biomass method obtains biomass data through field exploration. Standard sample plots are established based on forest type, and tree samples of various organs are harvested. The dry mass of these samples is then measured to determine the average biomass of the standard plot. By multiplying the average biomass by the number of plants, the biomass per unit area, or standard plot biomass, is calculated. Ultimately, carbon sequestration is determined by multiplying the standard plot biomass by the carbon content coefficient. For example, Liu et al. [57] used this method to estimate the aboveground biomass of karst forest stands in central Guizhou Province. However, direct-harvest method requires a large amount of manual labour and resources. It poses irreversible damage to plants and habitats and is not suitable for some fragile ecosystems [58]. While the data obtained through this method are accurate, it is challenging to scale up for large-scale and long-term studies, and it fails to capture the dynamic changes in the continuous carbon sink capacity of forests.

#### *3.1.2 Allometric equations and growth models*

The growth models are idealized equations that describe the relationship between biomass growth and various plant characteristics, based on the principles of relative growth theory. The allometric scaling models have been established based on the correlation between individual biomass and tree diameter at breast height (DBH) and/or tree height, and they have been widely used in biomass estimation. The biomass of the sample plot was calculated by sampling and felling a small number of representative standard trees within the plot. The existing allometric relationships between measurement indicators and biomass can be expressed by allometric equations:

$$W = a \times D^b \text{ and } W = a \times \left(D^2 \times H\right)^b \tag{1}$$

where, $W$ represents the biomass of various organs in the forest, $D$ represents the average diameter at breast height of the forest, $H$ represents the average diameter height, and $a$ and $b$ are parameters. Currently, relative growth models for biomass have been well-established for different vegetation types, providing a solid foundation for expanding forest carbon sequestration calculation methods [59, 60].

#### *3.1.3 Other forest carbon sequestration measurement methods based on biomass data*

With the improvement of biomass data and biomass relative growth models, and the increasing demand for large-scale forest carbon sequestration measurement, biomass expansion factor methods, biomass inventory methods, and accumulation volume expansion methods based on basic biomass data have gradually emerged and been widely applied.

The Biomass Expansion Factor (BEF) method calculates forest carbon sequestration using the ratio of stand biomass to wood volume as the BEF. It uses the average wood density to multiply the total volume to obtain the total biomass of certain forest types. This method is commonly used to estimate forest biomass at a national scale. The BEF value is the core of this method, typically derived from the expansion factor constant proposed by the Intergovernmental Panel on Climate Change (IPCC) or calculated using a continuous function based on forest type and age. Wu et al. [61] estimated carbon stocks of urban greening trees and discussed the change of BEF values in an urban environment. However, it should be noted that the BEF method ignores the contributions of understory vegetation and soil when estimating biomass. As large-scale forest carbon sink estimation research has become more prevalent at the national and even global scales, the carbon pools of understory vegetation, soil, and roots have garnered increased attention. Consequently, methods such as the Biomass Inventory and Accumulative Volume Expansion methods have gradually gained popularity.

The Biomass Inventory method is based on the relationship between forest stock and biomass and calculates forest carbon sequestration by combining ecological survey data and forest census data. It calculates the carbon storage density of the tree layer in each forest ecosystem type, and then estimates the total carbon storage per unit area of each forest type based on the ratio of tree layer biomass to total biomass. The biomass inventory method is simple and accurate, making it widely applied, especially suitable for long-term and large-scale forest carbon storage and sink monitoring. However, there are shortcomings such as high labour demand, indirect recording of carbon storage, inability to reflect dynamic effects such as seasonal and temporal changes, and neglect of soil carbon sequestration capacity.

The Accumulative Volume Expansion method, which is capable of measuring understory biomass has been proposed. This method takes the accumulation expansion coefficient as the core and calculates the total biomass based on forest accumulation. It specifically consists of three parts: tree biomass carbon sequestration, understory plant carbon sequestration, and forest carbon sequestration. The Accumulative Volume Expansion method is a continuation of the traditional biomass expansion factor method, which fully considers the carbon sequestration of tree species, understory, and forest land, and has the advantage of simple operation. However, the coefficients employed are based on average data from sample plot statistics, which may result in significant errors in the statistical results. Furthermore, this method does not account for factors such as tree age and organic carbon decomposition by soil microorganisms.

### 3.2 Remote sensing: Combination of remote sensing products and field data for regional estimates

With advancements in satellite communication technology, remote sensing has emerged as a powerful tool for acquiring large-scale vegetation data. Consequently, methods for estimating biomass based on remote sensing data have gained prominence. By combining these data with field observations, researchers can analyze the spatial distribution and temporal dynamics of vegetation, calculate the biomass distribution in forest ecosystems, and ultimately derive the spatiotemporal distribution of carbon, enabling the estimation of carbon sequestration in extensive forest areas.

### *3.2.1 Data sources of remote sensing*

Remote sensing data can be divided into optical remote sensing data, microwave radar data, and Light laser detection and ranging (LiDAR) data. Optical remote sensing data is the earliest remote sensing data source used for biomass estimation and is obtained through optical remote sensing technology. It remains a key data source for estimating large-scale forest biomass at national or global scales. This spectral passive remote sensing technology captures the reflectance spectral characteristics of vegetation. By measuring the relationship curve between reflectance spectral characteristics and vegetation category, chlorophyll content, and growth status, the condition and dynamic changes of vegetation can be monitored. Optical remote sensing data has the advantages of low cost and detailed provision of forest horizontal structure parameters, but data distortion is easily caused by saturation of remote sensing signals in areas with high biomass density. Meanwhile, there are shortcomings such as susceptibility to climate conditions and inability to accurately obtain vertical structure information of forest vegetation. Commonly used optical remote sensing datasets include the NOAA/Advanced very high-resolution radiometer (NOVAA/AVHRR), Landsat/Geomatic Mapper (Landsat/TM), and so on.

Microwave radar data, derived from active scanning by microwave satellites, is less affected by weather and cloud cover and offers better insights into forest conditions due to its longer wavelength. However, it can also suffer from signal saturation in high-biomass areas, and terrain can distort data when measuring forest vertical structure parameters. Currently, synthetic aperture radar (SAR) and polarimetric SAR (POLSAR) are the primary sources of microwave radar data.

LiDAR data, obtained through active scanning using LiDAR technology, can be acquired *via* airborne or spaceborne LiDAR. Airborne LiDAR provides precise canopy height data but has high requirements for sensors, flight altitude, and coverage, leading to high data acquisition costs and limited suitability for large-scale monitoring of forest vertical structure. Spaceborne LiDAR can depict spatial structure information across large forest canopies, with the potential for global data acquisition. However, its application on a large scale is constrained by insufficient spatial sampling continuity. It entails high data acquisition costs and demanding image processing techniques, and it may not fully capture the horizontal structure of forests.

Given the limitations of a single-source remote sensing data in estimating forest sequestration, the multiple-source data has become the fundamental data source for accurate forest carbon sequestration estimation. For example, Feng et al. [62] quantified carbon stock and its driving mechanism based on multi-source remote sensing data of urban forest in China.

### *3.2.2 Statistical measurement and AI methods based on remote sensing data*

Statistical measurement and artificial intelligence methods are main methods to estimate the forest carbon sequestration based on remote sensing data. The commonly used statistical measurement methods are regression analysis, while artificial intelligence methods mainly include Artificial Neural Network (ANN), Random Forest (RF), and so forth.

The Regression analysis method mainly uses forest aboveground biomass as the dependent variable and remote sensing spectral information, vegetation index, and texture features as independent variables to estimate the forest aboveground biomass and calculates carbon sequestration based on the carbon content coefficient. This approach is straightforward, easy to interpret, and relies on minimal data processing, making it an accessible and efficient tool for ecological studies.

The relationship between forest biomass and related factors often presents non-linear or even more complex forms. Traditional linear regression-based estimation of forest biomass often leads to significant estimation errors. With the development of machine learning technology, AI methods have greatly improved the accuracy of forest biomass estimation.

The ANN models can approach the objective function without relying on existing mathematical models through extensive training and simulation. Although the accuracy of ANN model is much higher than that of traditional regression methods, it cannot explain the internal mechanism of the model. Moreover, setting learning parameters require a lot of effort and time.

The RF is an algorithm model with basic units namely decision trees. Initially, the Bootstrap resampling method is employed to extract multiple samples, and decision trees are modeled for each sample, with each tree serving as a classifier. Subsequently, the prediction results from multiple decision trees (the forest) are aggregated, and the most frequently selected category is designated as the output result. Essentially, this constitutes an ensemble learning approach. RF is particularly suited for multivariate large sample datasets and can mitigate the risks of overfitting and multicollinearity. Li et al. [63] utilized RF and ANN models to estimate NPP of grasslands in the permafrost region of the Qinghai-Tibet Plateau. The NPP values obtained through these two algorithms exhibited similar accuracy and a high correlation with measured values, indicating the reliability of the results.

In general, these machine learning methods are advantageous for better characterizing and leveraging the logical relationships within data, fully exploiting the information embedded in vast datasets, and enhancing the accuracy of forest biomass estimations. However, these methods necessitate a high degree of relevance to the applicability of the data and research objectives, as well as a substantial amount of data in the learning datasets, making them particularly suitable for estimating forest biomass in the context of big data.

### *3.2.3 Empirical and mechanism models based on remote sensing data*

With the development of remote sensing technology and the improvement of remote sensing data, model simulation methods that integrate sample plot data, inventory data, and remote sensing data have rapidly developed, forming remote sensing model methods, such as the Miami model, and the Carnegie Ames Stanford Applech (CASA) model.

The Miami model, an empirical method, estimates forest ecosystem productivity (Net Primary Productivity, or NPP) by correlating it with annual precipitation and average temperature [64]. However, practical applications have revealed that NPP is influenced by factors beyond precipitation and temperature, including soil nutrients, light, and human activities. Consequently, current research typically does not directly apply the Miami model for NPP calculations. Instead, it is considered as a representation of potential NPP under climatic influence, or the model is refined by incorporating additional environmental variables. For example, Sha et al. [65] used the Miami model to calculate the potential NPP of grassland ecosystems in Inner Mongolia, compared it with measured values, and evaluated the impact of human activities on grassland carbon sequestration while eliminating climate factor interference.

Due to the oversimplification of the Miami model, the later Thomthwaite-Memorial model introduced actual evapotranspiration parameters for correction, while the Chikugo model, based on principles related to plant physiology and ecology, introduced parameters such as net radiation and radiation dryness and changed the model to a semiempirical model [66], which can achieve more accurate estimation results than the Miami model. Nevertheless, due to only considering climate factors such as solar radiation, evaporation, temperature, and moisture, the calculation results of the Chikugo model are considered potential NPP without taking into account key limiting factors such as soil and human interference [67].

The CASA stands as the most prevalent mechanism model based on light use efficiency [68], and it estimates forest NPP by photosynthetic active radiation absorbed by plants. This is achieved by multiplying APAR (absorbed photosynthetically active radiation) by light use efficiency, with APAR estimable through remote sensing:

$$\mathrm{NPP}(x, t) = \mathrm{APAR}(x, t) \times \varepsilon(x, t) \tag{2}$$

where, $\varepsilon$ is the light use efficiency. To enhance the CASA model's applicability, numerous studies have refined it by employing different maximum light use efficiency values for various vegetation types.

Other widely adopted models for light energy utilization include C-FIX and GLO-PEM [69]. Remote sensing-based modeling enables precise, real-time simulation and mapping of NPP distributions. This approach leverages existing multi-source data, emphasizes various ecosystem processes supported by data interoperability, and yields more reliable outcomes. While empirical and mechanism models are straightforward and require few parameters, they fall short in explaining the underlying physiological and ecological processes.

### 3.3 Eddy covariance: Measurement of forest carbon exchange

The Eddy Covariance method (EC) directly measures the net $CO_2$ exchange between terrestrial ecosystems and atmosphere within a footprint (usually several square-meters to several square-kilometers) based on micro-meteorological principles. Based on this, the regional scale net ecosystem productivity is estimated through scaling [70, 71]. The calculation formula of EC can be obtained:

$$F_C = \overline{d'w'} \tag{3}$$

where, $F_C$ is the carbon dioxide flux, $d'$ is the carbon dioxide concentration, and $w'$ is the wind speed in the vertical direction (both take the average fluctuation of their respective values in the vertical direction for a period of 15–30 minutes).

Key instruments required for EC includes a three-dimensional ultrasonic anemometer, an open/closed circuit infrared $CO_2/H_2O$ gas analyzer, and measuring instruments such as thermometer and hygrometer. Extensive research has employed EC method to quantify the carbon balance of forest ecosystems [72, 73] and used it for ecosystem model calibration and validation [74].

The EC method can achieve long-term continuous positioning observation of carbon flux on a fine time scale (such as every 30 minutes), which can reflect the impact of climate fluctuations on NEP. However, there are some limitations in the applications of EC, mainly including: (1) Flux observation stations are often set up in areas with less human interference, making it difficult to balance differences in forest age and ecosystem heterogeneity, resulting in biases in regional scale carbon sink inference results; (2) The measured carbon flux usually does not include the impact of disturbance factors such as logging and fires, which may overestimate the ecosystem carbon sink at the regional scale; (3) The EC method is mainly based on the principles of micro-meteorology. Due to factors such as missing observations, complex underlying surfaces and meteorological conditions, energy budget closure, and errors in observation instrument systems, it can cause observation errors and representativeness errors in carbon flux estimation.

### 3.4 Processed-based models: Simulation of forest carbon dynamics and future predictions

#### *3.4.1 Forest ecosystem processed-based models*

Processed-based models are mainly used to describe ecosystem processes or simulate the dependence of water and carbon cycles on interactive processes such as photosynthesis, transpiration, and respiration [75], including the Biogeochemical Cycles (Biome-BGC), Soil Carbon and Nitrogen Model (CENTURY), and so on.

The BIOME-BGC model is a biogeochemical model of vegetation, litter, soil moisture, storage, and flux of carbon and nitrogen at multi-scale and is evolved from the forest dynamics model. It is widely used in estimation of NPP in various forest ecosystems. Li et al. [63] used the Biome-BGC model to simulate NPP dynamics in broad-leaved Korean pine forests in Changbai Mountain, revealing an increasing trend in forest carbon sequestration with tree age. The BIOME-BGC model's flexibility in parameter adjustment for specific ecosystems allows it to represent short-term carbon flux dynamics and simulate long-term carbon sequestration patterns and their response to environmental changes [76]. It should be noted that the appropriate selection and modification of model parameters can have an impact on the accuracy of estimating NPP.

The CENTURY model, based on soil structure, simulates the biogeochemical cycling of carbon, nitrogen, and phosphorus. It integrates temperature and precipitation as driving factors to simulate and predict ecosystem productivity. In general, process-based models feature a rigorous structure, high accuracy, and a strong mechanistic foundation, offering biological interpretability. However, they require numerous parameters, which can be challenging to obtain completely [77]. Ecosystems' material and energy cycles and transformations are complex, and ecological processes vary between ecosystems, potentially leading to significant deviations. Additionally, productivity research with different backgrounds and objectives may focus on different ecological processes, resulting in various productivity estimates.

Processed-based models require a large amount of detailed ground information as input data and need to adjust parameters based on spatiotemporal differences; these will result in great uncertainties, especially in the regional application [75].

#### *3.4.2 Processed-based coupling models*

The processed-based coupling models integrates processed-based models with other models, eliminating their disadvantages in parameter acquisition and estimation error while ensuring the model mechanism, including The Atmosphere Vegetation Interaction Model (AVIM), GLOPEM-CEVSA model, and so forth. The

AVIM couples land surface physical processes with vegetation physiological and ecological processes, simulating the physical exchange of heat and water between vegetation and habitat, as well as physiological exchange processes such as photosynthesis and respiration. The GLOPEM-CEVSA model couples CEVSA processed-based model with the global production efficiency model (GLO-PEM). The model calculates the photosynthetically active radiation absorbed by vegetation, subtracts the vegetation respiration simulated by the respiration coefficient, and obtains forest carbon sequestration information. Zhang et al. [78] developed a GLOPEM-CEVSA model to estimate NPP in Chongqing and then analyzed the spatiotemporal distribution of NPP for different vegetation types.

Process-based coupling models merge the advantages of processed-based models and other models. Based on plant growth's physiological and biochemical principles, they offer biological interpretability, ensuring model stability and applicability. Furthermore, they simplify the number of parameters and are becoming a crucial tool for NPP estimation and prediction.

### 3.5 Deep learning models: Forest carbon estimation that integrating big data and temporal-spatial mapping

Deep learning, a branch of machine learning, combines numerous simple nonlinear modules to transform features into higher-level abstract features. It automatically extracts effective features from data for prediction [79]. With the advancement of remote sensing technology and the surge in ground observation data, ecological remote sensing has entered the big data era. Deep learning has made progress in simulating terrestrial ecosystems' carbon and water cycles. For example, Ben Abbes et al. [80] estimated soil moisture in Canadian agriculture using the Long Short Term Memory (LSTM) method, achieving good performance and accuracy. Wu et al. [81] simulated global forest NPP using a deep convolution model, reflecting key influencing factors for different forest types. Kaba et al. [82] estimated daily solar radiation in Türkiye using Deep Neural Network (DNN), producing accurate and comparable results. However, deep learning models still lack corresponding applications and research in other carbon flux simulation aspects.

Convolutional Neural Networks (CNN) are commonly used deep learning algorithms that have good processing capabilities for problems with large amounts of information, deeper data mining, and can more efficiently extract feature patterns. By hierarchically extracting spatial and temporal feature patterns, CNNs excel at deep data mining and efficiently capturing complex relationships between inputs and outputs. These models achieve high prediction accuracy by learning intricate mapping. Studies indicate that the deeper the computational depth, the more advantageous it is for classification and prediction [83], particularly accurate and efficient in solving multi-dimensional ecological data.

### 3.6 Integrated monitoring and prediction system for forest carbon sink

The quantification of forest ecosystem carbon sequestration is an important scientific basis for formulating policies to carbon emission mitigation and sink enhancement. The continuous improvement of carbon sink estimation methods has brought abundant data sources, creating possibilities for the accuracy of estimation and the formulation of subsequent countermeasures. The inconsistency of forest carbon sequestration estimation methods will bring great uncertainty to the estimation results, and need to take measures to establish an integrated monitoring and prediction system for forest carbon sink. It is necessary to construct an integrated forest ecosystem carbon balance measurement system based on multiple methods and multi-source data, from the perspectives of multiple processes and scales, that integrates "sky-air-ground".

For a long time, a large amount of observation and research work has been carried out on the carbon balance of forest ecosystems, laying a solid foundation for exploring the laws of forest carbon sinks and understanding the potential of forest carbon sinks. However, it is necessary to strengthen the observation of key areas of forest carbon sinks.

Firstly, it is necessary to strengthen the observation of underground carbon pools, especially changes in soil carbon pools because regular observation of soil carbon pools with unified standards is still insufficient [84, 85]. Extensive research has focused on the observation of aboveground carbon pools in forest ecosystems and can be combined with remote sensing methods to reflect changes in vegetation carbon pools at different spatiotemporal scales. In addition, it is necessary to strengthen the observation of carbon sinks in ecosystems affected by human interference because traditional ecosystem observations are usually established in ecosystems with less human interference. However, for the carbon sequestration function of forests, human activities, including human interference and ecosystem management, have an undeniable impact. The role of human activities in the national scale terrestrial ecosystem carbon sink still needs to be systematically investigated; Especially, there is still a lack of observation on the horizontal transfer and changes of organic carbon caused by forest logging, livestock grazing, and so on, which urgently needs to be strengthened.

The process-based ecosystem carbon cycle model is an important tool for predicting changes in forest carbon sinks. Many carbon cycle models can effectively simulate the response of natural ecosystems' carbon cycle processes to climate change [86, 87], but it is difficult to demonstrate the impact of human activities on ecosystems [88]. For example, only a few models can characterize the impact of forest management on ecosystem carbon sources and sinks [12]. Most models do not consider the lateral transport of organic carbon in terrestrial ecosystems [89] and the impact of forest age on carbon sinks [90]. Therefore, accurately quantifying the contributions of various components of natural and human activities to the potential for enhancing forest ecosystems remains a great challenge. Considering that China's terrestrial ecosystem carbon sink has largely benefited from ecological projects such as returning farmland to forests and afforestation in the past few decades [91], it is crucial to construct a natural and human coupled ecosystem carbon cycle process model for accurately predicting the carbon sink potential of forest ecosystems and quantitatively distinguishing the contributions of natural and human factors.

In addition, the model is difficult to simulate the response of forest ecosystem carbon sinks to extreme climate events [92]. Therefore, strengthening the prediction of forest carbon sink stability is crucial for accurately predicting changes in forest carbon sinks and enhancing ecosystem management. It is necessary to establish a carbon cycle parameter dataset suitable for forest ecosystems in the natural and human coupled ecosystem process model, considering the ecosystem's response to extreme climate events and other processes, to improve the accuracy and spatiotemporal resolution of model simulation.

## 4. Sustainable forest management utilization strategies

### 4.1 Forest management for enhancing carbon sinks: Afforestation, reforestation, conservation, and sustainable management

Forests are the largest and most stable biomass and organic carbon pool in the world. In the 2010s, global forest biomass carbon accumulated at a rate of 0.50 ± 0.20 Pg C yr.$^{-1}$, highlighting the importance of forests in climate mitigation strategies [6]. Through effective forest management activities such as afforestation, reforestation, conservation, and sustainable management, the carbon sink function of forests can be significantly maintained or further enhanced, thereby combating global climate change.

Afforestation and reforestation remain the natural-based solution for climate change mitigation. Natural ecosystems beyond existing forests, agriculture, and urban land could support an additional 900 million hectares of continuous forest. This would mean an increase in forest area of more than 25% and a cumulative carbon storage of 200 Pg C at maturity [93]. This value is astonishing, equivalent to approximately 70% of the total atmospheric carbon concentration attributed to human activities. However, there may not be sufficient land supporting for such extensive af/reforestation. A 60% increase in global forest cover could potentially lead to a more than doubling of food prices due to land conflicts against food production [94]. Researchers argue that the 200 Pg C estimate is excessive, suggesting that the actual carbon sequestration potential of af/reforestation may be less than half of 200 Pg C [95]. Between 1990 and 2020, the global area of planted forests increased by 123 million hectares, but the annual growth rate declined from 5.13 million hectares during 2000–2010 to 4.06 million hectares during 2010–2020 [96]. In addition to the impact on the carbon cycle, forests may also influence the climate through biophysical mechanisms. The decrease in albedo (reflectivity) resulting from af/reforestation (as forests have a darker color than soils, snow, and low vegetation) could offset a substantial portion of the carbon sequestration benefit and even contribute to local warming. In high-latitude regions covered by snow and ice or in mid-latitude arid regions dominated by sand or saline-alkali soils, where albedo is typically high, afforestation activities can disrupt highly reflective surfaces, leading to a warming effect that may even counteract the carbon sequestration benefits [97, 98].

Compared to af/reforestation, the conservation and sustainable management of existing forests incur lower costs and do not change the original land types. Therefore, they do not pose competition issues with other land uses, such as agricultural land [99]. On the other hand, natural forest ecosystems possess unique functioning that support higher biodiversity and provide additional benefits, including soil conservation, mitigation of air, and water pollution, among others [99]. Newly established forests inevitably alter the original community composition and structure. While monocultures of fast-growing tree species can effectively sequester carbon, they are weak in the ecosystem services and functioning that depend on biodiversity, which are inherently present in natural forests. Furthermore, as forests age, their carbon sequestration capacity may gradually decline, ultimately leading to a continuous reduction in carbon sequestration [100] and even a potential transition into carbon sources. Consequently, the sustainability of forest carbon sequestration, the challenges in quantifying changes in carbon stocks, and the environmental and socio-economic impacts of large-scale af/reforestation programs limit the adoption of forestry activities in climate mitigation strategies [101]. Determining how to manage forests, where to plant, how to plant, and which tree species to plant will become the primary focus of future efforts to enhance the climatic contributions of forestry.

### 4.2 Carbon resource utilization: Wood products as sustainable carbon storage

Land-use changes account for 10% of total emissions (1.0 ± 0.7 Gt C $yr.^{-1}$) [1], with deforestation being a primary driver of vegetation and soil carbon loss beyond agricultural activities. Over two-thirds of the carbon sink from forests has been negated by tropical deforestation [2]. Most felled forests are used for producing wood products, paper, or directly as fuel. All these are harvested wood products (HWP), with a specialized definition provided by UNFCCC as "wood-based materials harvested from forests, which are used for the production of commodities such as furniture, plywood, and paper and paper-like products, or for energy, and other fiber products from non-timber sources, such as rattan or bamboo, could also be considered wood products." [102].

HWPs not only provide economic value but also serve as crucial means for achieving carbon neutrality through storing carbon absorbed by forests and replacing fossil fuels [103]. HWPs are generally categorized into: Long-Lived Products (LLPs), including sawn wood, wood-based panels, and other industrial roundwood; Short-Lived Products (SLPs), such as paper and paperboard products; and Very Short-Lived Products (VSLPs), directly harvested or recycled waste from other wood product for energy. The lifetime of different HWPs varies significantly. For LLPs, especially construction products or furniture, their lifespan can exceed 40 years, but for SLPs or VSLPs such as paper and firewood, their life span is only 1–2 years [104]. LLPs can serve as an efficient and stable carbon sink through long-term carbon storage. Compared with other carbon sequestration technologies such as carbon capture and storage (CCS) and carbon capture, utilization, and storage (CCUS), there are no technical barriers or potential leakage risks, and it also promotes sustainable management of forests. Through rational logging and afforestation, the health and stability of forest ecosystems can be maintained, supporting multiple ecological functioning and services such as biodiversity conservation and soil and water conservation [105]. Therefore, LLPs offer significant climate benefits, whereas SLPs and VSLPs often serve as sources of greenhouse gas emissions.

The carbon stored in HWPs is becoming an important tool for mitigating climate change in many countries and regions. Studies indicated that the current global net carbon sequestration of HWPs amounts to 335 million tons of $CO_2$ equivalent ($CO_2e$) per year (2015), projected to reach 441 million tons of $CO_2e$ per year by 2030 [106]. This is equivalent to offsetting emissions under certain scenarios. The demand for HWPs is expected to continue growing in most countries by 2060, coinciding with the majority of countries "carbon neutrality" or "net-zero targets" set before 2060. Thus, at a national or regional scale, HWPs play a vital role as a "buffering agent" against climate change. However, caution is needed as the carbon sequestration provided by HWPs may significantly decline or even fail to offset historical emissions after 2060, due to decline demand for sawn wood and paper [107]. The increasing use of HWPs may also impact agricultural productivity, exacerbate land competition, and cause food price fluctuations. Increased demand for wood leading to deforestation can result in environmental degradation and soil erosion, subsequently reducing local agricultural productivity. Furthermore, af/reforestation action for producing HWPs may intensify competition for agricultural land, leading to higher food prices and an increased risk of hunger as mentioned in Section 4.1.

### 4.3 Renewable energy integration: Synergies with renewable energy projects

#### *4.3.1 Forest biomass energy*

Forest biomass used for energy or biofuels can be sourced directly from natural ecosystems. As a primary biomass source from forests, it includes organic products or residues derived directly from living or recently dead trees, which constituted nearly half of the world's harvested forest biomass [108]. Forest residues, comprising branches, treetops, and stumps, result from forest production and forest management, as well as bioenergy-specific woody crops. One prominent form of forest biomass energy is bioenergy from forest residues. The residue from forest tending and thinning constitutes the primary type of forest residue. For example, in China, tending and thinning residues account for 87% of the national total amount (133.05 million t; [109]). The common method for dealing with these forest residues is on-site burning rather than leaving them to decompose naturally because of occupying land and posing a risk of forest fire [110]. On-site burning leads to a series of issues such as air pollution and reduced land productivity [111]. Reusing the forest residue as forest bioenergy as a substitute for fossil fuels in power generation is considered a viable option to reduce GHG emissions in the power system, with biomass-derived electricity capable of reducing greenhouse gas emissions by 86–93% [112].

In addition to residue from forest production and management, specific woody crops are another source of forest biomass for energy. Short-rotation (3–15 years) techniques for growing species such as poplar (*Populus*), willow (*Salix*), eucalyptus/gum trees (*Eucalyptus*), and perennial grasses (*Miscanthus*), have been widely used over the past 2–3 decades. The sustainable management of forests can provide a significant amount of biomass for energy production, contributing to both energy security and climate change mitigation. By 2050, energy crops will surpass firewood as the primary source of biomass energy, contributing approximately 20% of the global energy supply, with wood crops playing a major role in this contribution [113]. Many agricultural countries and regions, such as the United States, promote short-rotation woody crops as they can serve as feedstocks for bioenergy, provide farmers with a stable income stream, and simultaneously avoid frequent soil disturbance while enhancing soil health, including improvements in soil structure, microbial lifespan, nutrient density, and carbon levels [114].

The integration of forest biomass energy into global energy systems holds significant potential for future development. As the demand for renewable energy continues to grow, the use of forest residues will become increasingly important. These sources of bioenergy not only contribute to energy security and climate mitigation but also provide economic opportunities for rural communities. However, ensuring sustainable harvesting of forest residues will also emerge as a key focus in future forest management. Excessive extraction leads to reduced soil carbon input and loss of unique habitats, consequently resulting in soil degradation, decreased biodiversity, and disruption of ecosystem services [115, 116]. Balancing the utilization of forest residues with other purposes is crucial for avoiding conflicts and maximizing the overall benefits to ecosystems. Therefore, establishing and adhering to sustainable forestry solutions is of paramount importance to avert these adverse effects.

### 4.3.2 Coupling with other renewables

Deploying large-scale renewable installations requires vast areas. Large quantities of renewables were located in forests, and deploying them led to the loss of large amounts of forest area. It shows that over 9% of solar farms have conflicts with forests globally, with three-fourths of that located directly in forest areas [117]. For onshore wind energy, nearly 18% were located in the forests. It takes a higher land-use footprint (installed capacity per land area) of wind farms compared to fossil fuels and other renewables [118, 119]. Vegetation removal, turbine foundation, and road expansion resulted in large clearing areas of forests [120]. Furthermore, to decrease the potential wind weakness downstream, it requires a distance of 7–10 rotor diameters between turbines to maximize the power generation, which takes large-scale landscape areas [121].

The fast expansion of renewable energy infrastructure poses a great threat to the carbon sequestration ability of vegetation [122] as well as other ecosystem functioning and services [123]. In forest ecosystems particularly, this has manifested through decreased vegetation indices [124], habitat fragmentation [125], and biodiversity loss [126]. Whether renewable mitigation could offset the loss of carbon stock and sink from deforestation and how to couple forestry and renewable development will be a very important study field.

The integration of solar photovoltaic (PV) systems with forest conservation is crucial for balancing renewable energy development and ecological protection. A key method is site selection optimization. Remote sensing technology, geographic information systems (GIS), life cycle assessment can be integrated to determine the optimal location of solar PV installations to minimize the occupation of forest land. By comprehensively analyzing the potential of solar energy, land-use patterns, soil types, and slope conditions, solar PV projects could be directed to bare land or degraded cropland, thereby reducing pressure on forests [127]. Another way is to develop the "forest-photovoltaic" system, in which solar panels are installed on the upper part of forest land while maintaining photosynthetic activities in the lower part [128]. Compared with the concept of forest photovoltaics, Agriphotovoltaics (APV), which combines photovoltaic systems and agricultural land, was proposed as early as 1982, using available space under solar panels mainly for crop growth [129]. Balancing power generation and crop yield with APV can increase land productivity by 70–80%, as the shading provided by solar panels can reduce evaporation and improve water efficiency [130]. With the widespread use of APV, there have also been many technological innovations in the forest-photovoltaic field. For example, installing "solar trees" on the edge of forests or beside mountain roads not only preserves the mountain landscape but also reduces construction costs and minimizes the loss of forest carbon sinks [128].

The impact of wind energy on forest fragmentation is nearly inevitable and extends throughout its entire life cycle. On the one hand, by adopting turbines with greater height and longer blades, we can capture more wind energy at higher altitudes and improve land use efficiency (i.e., higher installed capacity per unit of land), thereby minimizing the disruption to forest habitats. Using narrow roads or temporary access routes during construction could limit the disturbance to forest ecosystems. Implementing forest management plans that prioritize biodiversity conservation and ecosystem services can mitigate the negative impacts of wind farms.

### 4.4 Policy frameworks: Existing policies and recommendations for improvement

In response to the aforementioned sustainable development strategies for forest carbon sinks, an increasing number of countries and regions have adopted corresponding policies to ensure their smooth implementation.

#### *4.4.1 REDD+ policy framework and support*

Developing countries established the "REDD+" framework to protect forests as part of the Paris Agreement. "REDD" stands for "Reducing emissions from deforestation and forest degradation in developing countries". The "+" stands for "additional forest-related activities that protect the climate, namely sustainable management of forests and the conservation and enhancement of forest carbon stocks".

The REDD+ program operates through a results-based financing mechanism, where developing countries are provided with payments for reducing deforestation and forest degradation. This provides a strong incentive for developing countries to implement climate change mitigation policies. For example, studies have shown that REDD+ projects have led to significant reductions in deforestation and forest degradation rates in tropical regions [131]. The REDD+ framework has also fostered international cooperation, with developed countries providing financial and technical support to developing countries to implement REDD+ strategies. Currently, a total of 60 developing countries have reported REDD+ activities to the UNFCCC, with 17 of these countries reporting a reduction of nearly 11.6 billion tons of GHG emissions, which is over twice the net GHG emissions of the United States [132]. Forests protected and discovered under the "REDD+" framework provide food, energy, shelter, income, and medicine to over 1.6 billion people globally.

#### *4.4.2 Other projects and initiatives*

In addition to REDD+, there are numerous other initiatives and af/reforestation projects aimed at addressing climate change and environmental degradation. These national projects and initiatives involve both developed and developing countries and cover a wide range of fields, including renewable energy, energy efficiency, sustainable agriculture, and conservation of biodiversity.

In 2006, the UN Environment Programme proposed the Billion Tree Campaign (BTC). At the WEF Annual Meeting in 2020, governments from multiple countries and representatives from 300 businesses jointly adopted the BTC initiative (2020–2030) aimed at promoting large-scale reforestation investments and enhancing the global capacity to address climate change, slow down biodiversity loss, and other global challenges. China's ecological restoration projects, such as the TNSP (more details in Chapter 2.2), the Natural Forest Protection Project (NFPP), and the Grain for Green Program (GGP), are characterized by their large scale (with the GGP covering 82% of the country's land area), long duration (the TNSP has spanned over 40 years), and significant benefits. These projects have set a commendable example for many countries. Not only have they improved the ecological environment, but they have also adjusted the structure of agriculture and animal husbandry industries, promoting the development of rural economies. Another 15 km wide and 8000 km long "Great Green Wall" has been built across the African continent, spanning from Mauritania to Djibouti, to halt the encroachment of the Sahara Desert [133].

In West Africa, the carbon sequestration resulting from forest restoration is equivalent to approximately 27% of the region's total GHG emissions and is projected to reach 30% by 2030 [134]. The "National Greening Program" launched by the Philippines in 2011 aimed to plant 1.5 billion trees on 1.5 million hectares of land to increase forest cover and enhance carbon sequestration capacity. Some countries implement afforestation initiatives within urban areas. For example, the Million Trees LA initiative intends to improve Los Angeles's environment through planting 1 million trees. A 35-year 1-million-tree planting project can result in direct carbon sequestration of 448,000 t $CO_2$ and reduce emissions from power plants associated with urban building energy use by 576,000 t $CO_2$ [135]. Additionally, it offers benefits such as esthetic enhancement, stormwater runoff reduction, energy savings, and air quality improvement.

Many af/reforestation projects and policies have not only addressed the issue of land degradation but also, to a certain extent, alleviated poverty, enhanced food security, and improved economic prospects for people. However, policies promoted at the national or regional level may, under the guise of "mitigation," conceal a series of issues such as infringement on land use, interspecies conflict, and water resource depletion, ultimately causing greater harm to fragile local ecosystems and the communities that depend on these [136].

## 5. Conclusion

Forests, as the largest carbon sink in terrestrial ecosystems, play an irreplaceable role in mitigating climate change. However, we are currently facing a series of challenges, such as deforestation, forest degradation, forest aging, and the need for more accurate quantification of forest carbon sequestration. To address these challenges, it is crucial to leverage advanced technologies for monitoring, accounting, and modeling carbon stocks and sinks. These technologies, combined with efficient forest management and strategic use of forest resources, are essential for maintaining the carbon sequestration function of global forests and thereby creating emission reduction spaces for other sectors. Moreover, it is imperative to fully tap into the multifaceted functions of forests. Beyond their carbon sequestration capabilities, forests also support biodiversity conservation and provide essential ecosystem services such as surface carbon storage, soil conservation, and water retention, which contribute to human well-being. Promoting sustainable forest utilization strategies, including the use of wooden products for carbon resource utilization, integrating renewable energy sources, and establishing supportive policy frameworks, is vital for enhancing forest carbon sequestration and ensuring the sustainable development of global forests.

*DOI: http://dx.doi.org/10.5772/intechopen.1009089*

## Author details

Kang Xu[1*], Guiwu Zou[2] and Hanjian Hu[3]

1 School of Environmental Science and Engineering, Wuxi University, Wuxi, China

2 Jiangxi Provincial Key Laboratory of Conservation Biology, College of Forestry, Jiangxi Agricultural University, Nanchang, China

3 Urban Construction College, Zhejiang Shuren University, Shaoxing, China

*Address all correspondence to: xukang@cwxu.edu.cn

## References


[1] Friedlingstein P, O'Sullivan M, Jones MW, Andrew RM, Hauck J, Landschützer P, et al. Global carbon budget 2024. Earth System Science Data Discussions. 2024;**2024**:1-133

[2] Pan Y, Birdsey RA, Phillips OL, Houghton RA, Fang J, Kauppi PE, et al. The enduring world forest carbon sink. Nature. 2024;**631**:563-569

[3] Baccini A, Walker W, Carvalho L, Farina M, Sulla-Menashe D, Houghton RA. Tropical forests are a net carbon source based on aboveground measurements of gain and loss. Science. 2017;**358**:230-234

[4] Dixon RK, Solomon AM, Brown S, Houghton RA, Trexier MC, Wisniewski J. Carbon pools and flux of global forest ecosystems. Science. 1994;**263**:185-190

[5] Nunes LJR, Meireles CIR, Pinto Gomes CJ, Almeida Ribeiro NMC. Forest contribution to climate change mitigation: Management oriented to carbon capture and storage. Climate. 2020;**8**:21

[6] Pan Y, Birdsey RA, Fang J, Houghton R, Kauppi PE, Kurz WA, et al. A large and persistent carbon sink in the World's forests. Science. 2011;**333**:988-993

[7] Grelle A, Hedwall P-O, Strömgren M, Håkansson C, Bergh J. From source to sink—Recovery of the carbon balance in young forests. Agricultural and Forest Meteorology. 2023;**330**:109290

[8] Fang J, Yu G, Liu L, Hu S, Chapin FS 3rd. Climate change, human impacts, and carbon sequestration in China. Proceedings of the National Academy of Sciences of the United States of America. 2018;**115**:4015-4020

[9] Trugman AT, Quetin GR. Leveraging uncertainty in terrestrial ecosystem carbon stocks and fluxes. Earth's Future. 2023;**11**:e2022EF003322

[10] Alaniz AJ, Carvajal MA, Marquet PA, Vergara PM, Meneses L, Moreira-Arce D. Analyzing the spatiotemporal patterns of forests carbon sink and sources between 2000 and 2019. Earth's Future. 2022;**10**:e2021EF002560

[11] Harris NL, Gibbs DA, Baccini A, Birdsey RA, de Bruin S, Farina M, et al. Global maps of twenty-first century forest carbon fluxes. Nature Climate Change. 2021;**11**:234-240

[12] Pugh TAM, Lindeskog M, Smith B, Poulter B, Arneth A, Haverd V, et al. Role of forest regrowth in global carbon sink dynamics. Proceedings of the National Academy of Sciences of the United States of America. 2019;**116**:4382-4387

[13] Zeng J, Matsunaga T, Tan ZH, Saigusa N, Shirai T, Tang Y, et al. Global terrestrial carbon fluxes of 1999-2019 estimated by upscaling eddy covariance data with a random forest. Scienfitic Data. 2020;**7**:313

[14] Schimel D, Stephens BB, Fisher JB. Effect of increasing $CO_2$ on the terrestrial carbon cycle. Proceedings of the National Academy of Sciences of the United States of America. 2015;**112**:436-441

[15] Yang Y, Shi Y, Sun W, Chang J, Zhu J, Chen L, et al. Terrestrial carbon sinks in China and around the world and their contribution to carbon neutrality. Science China Life Sciences. 2022;**65**:861-895

[16] Ritchie H. Forest area. Published online at OurWorldinData.org. 2021. Available from: https://ourworldindata.org/forest-area

[17] Walker WS, Gorelik SR, Cook-Patton SC, Baccini A, Farina MK, Solvik KK, et al. The global potential for increased storage of carbon on land. Proceedings of the National Academy of Sciences of the United States of America. 2022;**119**:e2111312119

[18] Yao L, Liu T, Qin J, Jiang H, Yang L, Smith P, et al. Carbon sequestration potential of tree planting in China. Nature Communications. 2024;**15**:8398

[19] Madrigal-Gonzalez J, Calatayud J, Ballesteros-Canovas JA, Escudero A, Cayuela L, Rueda M, et al. Climate reverses directionality in the richness-abundance relationship across the World's main forest biomes. Nature Communications. 2020;**11**:5635

[20] Scharlemann JPW, Tanner EVJ, Hiederer R, Kapos V. Global soil carbon: Understanding and managing the largest terrestrial carbon pool. Carbon Management. 2014;**5**:81-91

[21] Ameray A, Bergeron Y, Valeria O, Montoro Girona M, Cavard X. Forest carbon management: A review of silvicultural practices and management strategies across boreal, temperate and tropical forests. Current Forestry Reports. 2021;**7**:245-266

[22] Walker XJ, Baltzer JL, Cumming SG, Day NJ, Ebert C, Goetz S, et al. Increasing wildfires threaten historic carbon sink of boreal forest soils. Nature. 2019;**572**:520-523

[23] Wang JA, Baccini A, Farina M, Randerson JT, Friedl MA. Disturbance suppresses the aboveground carbon sink in North American boreal forests. Nature Climate Change. 2021;**11**:435-441

[24] Zhang XM, Brandt M, Yue YM, Tong XW, Wang KL, Fensholt R. The carbon sink potential of southern China after two decades of afforestation. Earth's Future. 2022;**10**:e2022EF002674

[25] Xu H, Yue C, Zhang Y, Liu D, Piao S. Forestation at the right time with the right species can generate persistent carbon benefits in China. Proceedings of the National Academy of Sciences of the United States of America. 2023;**120**:e2304988120

[26] Harrington RA, Fownes JH, Vitousek PM. Production and resource use efficiencies in N- and P-limited tropical forests: A comparison of responses to long-term fertilization. Ecosystems. 2001;**4**:646-657

[27] Yu Z, Dong Y, Lu C, Agathokleous E, Zhang L, Liu S, et al. China's forestation on marginal land was less efficient in carbon sequestration compared with non-marginal land. One Earth. 2023;**6**:1692-1702

[28] Håkansson C, Hedwall P-O, Bader MK-F, Strömgren M, Axelsson M, Bergh J. Forest fertilization transiently increases soil $CO_2$ efflux in young Norway spruce stands in Sweden. Agricultural and Forest Meteorology. 2025;**360**:110287

[29] Lee J, Zhou X, Lee ST, Yang Y, Yun J, Lee HH, et al. Thinning enhances forest soil C storage by shifting the soil toward an oligotrophic condition. Science of the Total Environment. 2024;**925**:171745

[30] Wang Z, Wang G, Li Y, Zhang Z. Determinants of carbon sequestration in thinned forests. Science of the Total Environment. 2024;**951**:175540

[31] Ammer C. Diversity and forest productivity in a changing climate. New Phytologist. 2019;**221**:50-66

[32] Huang Y, Chen Y, Castro-Izaguirre N, Baruffol M, Brezzi M,

Lang A, et al. Impacts of species richness on productivity in a large-scale subtropical forest experiment. Science. 2018;**362**:80-83

[33] Chen X, Taylor AR, Reich PB, Hisano M, Chen HYH, Chang SX. Tree diversity increases decadal forest soil carbon and nitrogen accrual. Nature. 2023;**618**:94-101

[34] Crockett ETH, Atkins JW, Guo Q, Sun G, Potter KM, Ollinger S, et al. Structural and species diversity explain aboveground carbon storage in forests across the United States: Evidence from GEDI and forest inventory data. Remote Sensing of Environment. 2023;**295**:113703

[35] Zhai L, Will RE, Zhang B. Structural diversity is better associated with forest productivity than species or functional diversity. Ecology. 2024;**105**:e4269

[36] Zhang W, Xi Y, Brandt M, Ren C, Bai J, Ma Q, et al. Stand structure of tropical forests is strongly associated with primary productivity. Communications Earth & Environment. 2024;**5**:796

[37] Ontl TA, Janowiak MK, Swanston CW, Daley J, Handler S, Cornett M, et al. Forest management for carbon sequestration and climate adaptation. Journal of Forestry. 2020;**118**:86-101

[38] Seddon N, Chausson A, Berry P, Girardin CAJ, Smith A, Turner B. Understanding the value and limits of nature-based solutions to climate change and other global challenges. Philosophical Transactions of the Royal Society of London. Series B, Biological Sciences. 2020;**375**:20190120

[39] Jay AK, Crimmins AR, Avery CW, Dahl TA, Dodder RS, Hamlington BD, et al. Overview: Understanding risks, impacts, and responses. In: Crimmins AR, Avery CW, Easterling DR, Kunkel KE, Stewart BC, Maycock TK, editors. Fifth National Climate Assessment. Washington, DC, USA: U.S. Global Change Research Program; 2023

[40] Girardin CAJ, Jenkins S, Seddon N, Allen M, Lewis SL, Wheeler CE, et al. Nature-based solutions can help cool the planet—If we act now. Nature. 2021;**593**:191-194

[41] Zhai J, Wang L, Liu Y, Wang C, Mao X. Assessing the effects of China's three-north shelter Forest program over 40 years. Science of the Total Environment. 2023;**857**:159354

[42] Zhu J, Sun Y, Zheng X, Yang K, Wang GG, Xia C, et al. A large carbon sink induced by the implementation of the largest afforestation program on earth. Ecological Processes. 2023;**12**:44

[43] Busch J, Ferretti-Gallon K. What drives deforestation and what stops it? A meta-analysis. Review of Environmental Economics and Policy. 2017;**11**:3-23

[44] Gatti LV, Basso LS, Miller JB, Gloor M, Gatti Domingues L, Cassol HLG, et al. Amazonia as a carbon source linked to deforestation and climate change. Nature. 2021;**595**:388-393

[45] Allen CD, Breshears DD, McDowell NG. On underestimation of global vulnerability to tree mortality and forest die-off from hotter drought in the Anthropocene. Ecosphere. 2015;**6**:129

[46] Bracalini M, Balacenoiu F, Panzavolta T. Forest health under climate change: Impact of insect pests. Iforest-Biogeosciences and Forestry. 2024;**17**:295-299

[47] Luo KW, Wang XL, de Jong M, Flannigan M. Drought triggers and

sustains overnight fires in North America. Nature. 2024;**627**:321-327

[48] Rakotonarivo OS, Rakotoarisoa M, Rajaonarivelo HM, Raharijaona S, Jones JPG, Hockley N. Resolving land tenure security is essential to deliver forest restoration. Communications Earth & Environment. 2023;**4**:179

[49] Gren I-M, Aklilu AZ. Policy design for forest carbon sequestration: A review of the literature. Forest Policy and Economics. 2016;**70**:128-136

[50] Lohman A, Mason S, Olander L. "A Menu of State Actions to Promote Forest Carbon Sequestration and Storage" NI PB 22-14. Durham, NC: Duke University; 2022. Available from: http://nicholasinstitute.duke.edu/publications

[51] Xu J. Estimation of the spatial distribution of potential forestation land and its climatic potential productivity in China. Acta Geographica Sinica. 2023;**78**:677-693. [in Chinese with English abstract]

[52] Upton B, Miner R, Spinney M, Heath LS. The greenhouse gas and energy impacts of using wood instead of alternatives in residential construction in the United States. Biomass & Bioenergy. 2008;**32**:1-10

[53] Zhu JJ, Gao T, Yu LZ, Yang K, Sun T, Lu DL, et al. Carbon sink of forest ecosystems: Concept, time effect and improvement approaches. Chinese Journal of Applied Ecology. 2024;**35**:2313-2321. [in Chinese with English abstract]

[54] Uddien Shaik R, Alipour M, Rowell E, Watts A, Woodall C, Taciroglu E. Remote sensing and mapping of fine Woody carbon with satellite imagery and super learner. IEEE Geoscience and Remote Sensing Letters. 2025;**22**:1-5

[55] Chen QD, Gao T, Zhu JJ, Wu FY, Li XF, Lu DL, et al. Individual tree segmentation and tree height estimation using leaf-off and leaf-on UAV-LiDAR data in dense deciduous forests. Remote Sensing. 2022;**14**:2787

[56] Zhu J, Geoff Wang G, Zhang H, Gao T. On the research of climate-smart forestry. Scientia Silvae Sinicae. 2024;**60**:1-7. [in Chinese with English abstract]

[57] Liu L, Zhou Y, Cheng A, Wang S, Cai X, Ni J. Aboveground biomass estimate of a karst forest in Central Guizhou Province, southwestern China based on direct harvest method. Acta Ecologica Sinica. 2020;**40**:4455-4461. [in Chinese with English abstract]

[58] Bai G, Koehler-Cole K, Scoby D, Thapa VR, Basche A, Ge YF. Enhancing estimation of cover crop biomass using field-based high-throughput phenotyping and machine learning models. Frontiers in Plant Science. 2024;**14**:1277672

[59] Cudjoe E, Bravo F, Ruiz-Peinado R. Allometry and biomass dynamics in temperate mixed and monospecific stands: Contrasting response of scots pine (*Pinus sylvestris* L.) and sessile oak (*Quercus petraea* (Matt.) Liebl.). Science of the Total Environment. 2024;**953**:176061

[60] Price CA, Branoff B, Cummins K, Kakai RG, Ogurcak D, Papes M, et al. Global data compilation across climate gradients supports the use of common allometric equations for three transatlantic mangrove species. Ecology and Evolution. 2024;**14**:e70577

[61] Wu LL, Shi YJ, Zhang FY, Zhou YF, Ding ZT, Lv SX, et al. Estimating carbon stocks and biomass expansion factors of urban greening trees using terrestrial laser scanning. Forests. 2022;**13**:1389

[62] Feng RD, Liu SH, Wang FY, Wang KY, Gao P, Xu LL. Quantifying the environmental synergistic effect of cooling-air purification-carbon sequestration from urban forest in China. Journal of Cleaner Production. 2024;**448**:141514

[63] Li XH, Yu DP, Dai LM, Sun JX. Changes of productivity with stand development in broadleaf-Korean pine forest in Changbai Mountain, China. Chinese Journal of Applied Ecology. 2020;**31**:706-716. [in Chinese with English abstract]

[64] Lieth H. Modeling the Primary Productivity of the World. In: Lieth H, Whittaker RH, editors. Primary Productivity of the Biosphere. Ecological Studies. Vol 14. Berlin: Springer; 1975. pp. 237-263

[65] Sha ZY, Bai YF, Lan H, Liu XF, Li RR, Xie YC. Can more carbon be captured by grasslands? A case study of Inner Mongolia, China. Science of the Total Environment. 2020;**723**:138085

[66] Ohta S, Uchijima Z, Oshima Y. Probable effects of $Co_2$-induced climatic changes on net primary productivity of terrestrial vegetation In East-Asia. Ecological Research. 1993;**8**:199-213

[67] Wu YY, Wu ZF, Yu SX. Quantitative assessment of the impacts of human activities on net primary productivity. Chinese Journal of Applied Ecology. 2017;**28**:2535-2544. [in Chinese with English abstract]

[68] Guo D, Song XN, Hu RH, Cai SH, Zhu XM, Hao YB. Grassland type-dependent spatiotemporal characteristics of productivity in Inner Mongolia and its response to climate factors. Science of the Total Environment. 2021;**775**:145644

[69] Wang CL, Jiang QO, Deng XZ, Lv KX, Zhang ZH. Spatio-temporal evolution, future trend and phenology regularity of net primary productivity of forests in Northeast China. Remote Sensing. 2020;**12**:3670

[70] Yao YT, Li ZJ, Wang T, Chen AP, Wang XH, Du MY, et al. A new estimation of China's net ecosystem productivity based on eddy covariance measurements and a model tree ensemble approach. Agricultural and Forest Meteorology. 2018;**253**:84-93

[71] Yu GR, Chen Z, Piao SL, Peng CH, Ciais P, Wang QF, et al. High carbon dioxide uptake by subtropical forest ecosystems in the east Asian monsoon region. Proceedings of the National Academy of Sciences of the United States of America. 2014;**111**:4910-4915

[72] Krejza J, Haeni M, Darenova E, Foltynovova L, Fajstavr M, Sve J, et al. Disentangling carbon uptake and allocation in the stems of a spruce forest. Environmental and Experimental Botany. 2022;**196**:104787

[73] Zhu SY, Clement R, McCalmont J, Davies CA, Hill T. Stable gap-filling for longer eddy covariance data gaps: A globally validated machine-learning approach for carbon dioxide, water, and energy fluxes. Agricultural and Forest Meteorology. 2022;**314**:108777

[74] Wei D, Qi YH, Ma YM, Wang XF, Ma WQ, Gao TG, et al. Plant uptake of $CO_2$ outpaces losses from permafrost and plant respiration on the Tibetan plateau. Proceedings of the National Academy of Sciences of the United States of America. 2021;**118**:e2015283118

[75] Su Y, Zhang WF, Liu BJ, Tian X, Chen SX, Wang HY, et al. Forest carbon flux simulation using multi-source data and incorporation of remotely sensed model with process-based model. Remote Sensing. 2022;**14**:4766

[76] Chen YR, Xiao WF. Estimation of Forest NPP and carbon sequestration in the three gorges reservoir area, using the Biome-BGC model. Forests. 2019;**10**:149

[77] Zhuang Q, He J, Lu Y, Ji L, Xiao J, Luo T. Carbon dynamics of terrestrial ecosystems on the Tibetan plateau during the 20th century: An analysis with a process-based biogeochemical model. Global Ecology and Biogeography. 2010;**19**:649-662

[78] Zhang R, Zhou Y, Luo HX, Wang FT, Wang SX. Estimation and analysis of spatiotemporal dynamics of the net primary productivity integrating efficiency model with process model in karst area. Remote Sensing. 2017;**9**:477

[79] LeCun Y, Bengio Y, Hinton G. Deep learning. Nature. 2015;**521**:436-444

[80] Ben Abbes A, Magagi R, Goita K, Ieee. Soil moisture estimation from smap observations using long short-term memory (LSTM). In: IEEE International Geoscience and Remote Sensing Symposium (IGARSS). Yokohama, Japan; 2019. pp. 1590-1593

[81] Wu WJ, Gong C, Li XW, Guo HD, Zhang L. An online deep convolutional model of gross primary productivity and net ecosystem exchange estimation for global forests. IEEE Journal of Selected Topics in Applied Earth Observations and Remote Sensing. 2019;**12**:5178-5188

[82] Kaba K, Sarigul M, Avci M, Kandirmaz HM. Estimation of daily global solar radiation using deep learning model. Energy. 2018;**162**:126-135

[83] Simonyan K, Zisserman A. Very deep convolutional networks for large-scale image recognition. Computer Science. 2014

[84] Ding JZ, Wang T, Piao SL, Smith P, Zhang GL, Yan ZJ, et al. The paleoclimatic footprint in the soil carbon stock of the Tibetan permafrost region. Nature Communications. 2019;**10**:4195

[85] Wang TH, Yang DW, Yang YT, Piao SL, Li X, Cheng GD, et al. Permafrost thawing puts the frozen carbon at risk over the Tibetan plateau. Science Advances. 2020;**6**:eaaz3513

[86] Fisher RA, Koven CD. Perspectives on the future of land surface models and the challenges of representing complex terrestrial systems. Journal of Advances in Modeling Earth Systems. 2020;**12**:e2018MS001453

[87] Friedlingstein P, O'Sullivan M, Jones MW, Andrew RM, Hauck J, Olsen A, et al. Global carbon budget 2020. Earth System Science Data. 2020;**12**:3269-3340

[88] Bonan GB, Doney SC. Climate, ecosystems, and planetary futures: The challenge to predict life in earth system models. Science. 2018;**359**:eaam8328

[89] Ciais P, Yao YT, Gasser T, Baccini A, Wang YL, Lauerwald R, et al. Empirical estimates of regional carbon budgets imply reduced global soil heterotrophic respiration. National Science Review. 2021;**8**:nwaa145

[90] Yao YT, Piao SL, Wang T. Future biomass carbon sequestration capacity of Chinese forests. Science Bulletin. 2018;**63**:1108-1117

[91] Lu F, Hu HF, Sun WJ, Zhu JJ, Liu GB, Zhou WM, et al. Effects of national ecological restoration projects on carbon sequestration in China from 2001 to 2010. Proceedings of the National Academy of Sciences of the United States of America. 2018;**115**:4039-4044

[92] Schewe J, Gosling SN, Reyer C, Zhao F, Ciais P, Elliott J, et al. State-of-the-art global models underestimate

impacts from climate extremes. Nature Communications. 2019;**10**:1005

[93] Bastin J-F, Finegold Y, Garcia C, Mollicone D, Rezende M, Routh D, et al. The global tree restoration potential. Science. 2019;**365**:76-79

[94] Kreidenweis U, Humpenöder F, Stevanović M, Bodirsky BL, Kriegler E, Lotze-Campen H, et al. Afforestation to mitigate climate change: Impacts on food prices under consideration of albedo effects. Environmental Research Letters. 2016;**11**:085001

[95] Taylor SD, Marconi S. Rethinking global carbon storage potential of trees. A comment on Bastin et al. (2019). Annals of Forest Science. 2020;**77**:23

[96] FAO. Global Forest Resources Assessment 2020: Main report. Rome; 2020

[97] Canadell JG, Raupach MR. Managing forests for climate change mitigation. Science. 2008;**320**:1456-1457

[98] Rohatyn S, Yakir D, Rotenberg E, Carmel Y. Limited climate change mitigation potential through forestation of the vast dryland regions. Science. 2022;**377**:1436-1439

[99] Hua F, Bruijnzeel LA, Meli P, Martin PA, Zhang J, Nakagawa S, et al. The biodiversity and ecosystem service contributions and trade-offs of forest restoration approaches. Science. 2022;**376**:839-844

[100] Leng Y, Li W, Ciais P, Sun M, Zhu L, Yue C, et al. Forest aging limits future carbon sink in China. One Earth. 2024;**7**:822-834

[101] Zeng Y, Sarira TV, Carrasco LR, Chong KY, Friess DA, Lee JSH, et al. Economic and social constraints on reforestation for climate mitigation in Southeast Asia. Nature Climate Change. 2020;**10**:842-844

[102] UNFCCC. Estimation, reporting and accounting of harvested wood products. Technical paper; 2003

[103] Soimakallio S, Kalliokoski T, Lehtonen A, Salminen O. On the trade-offs and synergies between forest carbon sequestration and substitution. Mitigation and Adaptation Strategies for Global Change. 2021;**26**:4

[104] Lun F, Liu M, Zhang D, Li W, Liu J. Life cycle analysis of carbon flow and carbon footprint of harvested wood products of *Larix principis-rupprechtii* in China. Sustainability. 2016;**8**:247

[105] Wang M, Yue C, He J, Zhang P, Xu M, Li Y, et al. Can wood harvest followed by forest regrowth enhance carbon sequestration of the forest sector in China? Global and Planetary Change. 2024;**243**:104626

[106] Johnston CMT, Radeloff VC. Global mitigation potential of carbon stored in harvested wood products. Proceedings of the National Academy of Sciences. 2019;**116**:14526-14531

[107] FAO. Global Forest Products Facts and Figures 2023. Rome; 2024

[108] Malico I, Gonçalves AC. Sources and distribution of Forest biomass for energy. In: Gonçalves AC, Malico I, editors. Forest Bioenergy: From Wood Production to Energy Use. Cham: Springer International Publishing; 2024. pp. 25-52

[109] Tian J, Yu F, Guo Z, Yan Y, Xie Y, Ma Z, et al. Forest residues contribute significantly to climate change in China.

Resources, Conservation and Recycling. 2023;**198**:107157

[110] Puettmann M, Sahoo K, Wilson K, Oneil E. Life cycle assessment of biochar produced from forest residues using portable systems. Journal of Cleaner Production. 2020;**250**:119564

[111] Fann N, Alman B, Broome RA, Morgan GG, Johnston FH, Pouliot G, et al. The health impacts and economic value of wildland fire episodes in the U.S.: 2008-2012. Science of the Total Environment. 2018;**610-611**:802-809

[112] Xu H, Latta G, Lee U, Lewandrowski J, Wang M. Regionalized life cycle greenhouse gas emissions of Forest biomass use for electricity generation in the United States. Environmental Science & Technology. 2021;**55**:14806-14816

[113] Errera MR, Dias TADC, Maya DMY, Lora EES. Global bioenergy potentials projections for 2050. Biomass and Bioenergy. 2023;**170**:106721

[114] Ile OJ, McCormick H, Skrabacz S, Bhattacharya S, Aguilos M, Carvalho HDR, et al. Integrating short rotation woody crops into conventional agricultural practices in the Southeastern United States: A review. Land. 2022;**12**:10

[115] Pillay R, Watson JEM, Hansen AJ, Jantz PA, Aragon-Osejo J, Armenteras D, et al. Humid tropical vertebrates are at lower risk of extinction and population decline in forests with higher structural integrity. Nature Ecology & Evolution. 2022;**6**:1840-1849

[116] Qu X, Li X, Bardgett RD, Kuzyakov Y, Revillini D, Sonne C, et al. Deforestation impacts soil biodiversity and ecosystem services worldwide. Proceedings of the National Academy of Sciences of the United States of America. 2024;**121**:e2318475121

[117] Zhang P, Yue C, Li Y, Tang X, Liu B, Xu M, et al. Revisiting the land use conflicts between forests and solar farms through energy efficiency. Journal of Cleaner Production. 2024;**434**:139958

[118] Wachs E, Engel B. Land use for United States power generation: A critical review of existing metrics with suggestions for going forward. Renewable and Sustainable Energy Reviews. 2021;**143**:110911

[119] Wang C, Wang Y, Tong X, Ulgiati S, Liang S, Xu M, et al. Mapping potentials and bridging regional gaps of renewable resources in China. Renewable and Sustainable Energy Reviews. 2020;**134**:110337

[120] Diffendorfer JE, Dorning MA, Keen JR, Kramer LA, Taylor RV. Geographic context affects the landscape change and fragmentation caused by wind energy facilities. PeerJ. 2019;**7**:e7129

[121] Dupont E, Koppelaar R, Jeanmart H. Global available wind energy with physical and energy return on investment constraints. Applied Energy. 2018;**209**:322-338

[122] Zhang B, Zhang R, Li Y, Wang S, Xing F. Ignoring the effects of photovoltaic array deployment on greenhouse gas emissions may Lead to overestimation of the contribution of photovoltaic power generation to greenhouse gas reduction. Environmental Science & Technology. 2023;**57**:4241-4252

[123] Grodsky SM, Hernandez RR. Reduced ecosystem services of desert plants from ground-mounted solar energy development. Nature Sustainability. 2020;**3**:1036-1043

[124] Qin Y, Li Y, Xu R, Hou C, Armstrong A, Bach E, et al. Impacts of 319 wind farms on surface temperature and vegetation in the United States. Environmental Research Letters. 2022;**17**:024026

[125] UCS. Environmental Impacts of Renewable Energy Technologies; 2013

[126] Msigwa G, Ighalo JO, Yap PS. Considerations on environmental, economic, and energy impacts of wind energy generation: Projections towards sustainability initiatives. Science of the Total Environment. 2022;**849**:157755

[127] Choi CS, Ravi S, Siregar IZ, Dwiyanti FG, Macknick J, Elchinger M, et al. Combined land use of solar infrastructure and agriculture for socioeconomic and environmental co-benefits in the tropics. Renewable and Sustainable Energy Reviews. 2021;**151**:111610

[128] Um DB. Exploring the operational potential of the forest-photovoltaic utilizing the simulated solar tree. Scientific Reports. 2022;**12**:12838

[129] Goetzberger A, Zastrow A. On the coexistence of solar-energy conversion and plant cultivation. International Journal of Solar Energy. 1982;**1**:55-69

[130] Ghosh A. Nexus between agriculture and photovoltaics (agrivoltaics, agriphotovoltaics) for sustainable development goal: A review. Solar Energy. 2023;**266**:112146

[131] Guizar-Coutino A, Jones JPG, Balmford A, Carmenta R, Coomes DA. A global evaluation of the effectiveness of voluntary REDD+ projects at reducing deforestation and degradation in the moist tropics. Conservation Biology. 2022;**36**:e13970

[132] Change UC. A Decade of REDD+: Notable Achievements by Forest Nations; 2023

[133] McLymont R. Toward clean and green. Network Journal. 2010;**17**:20

[134] Buchanan GM, Field RH, Bradbury RB, Luraschi B, Vickery JA. The impact of tree loss on carbon management in West Africa. Carbon Management. 2021;**12**:623-633

[135] McPherson EG, Simpson JR, Xiao Q, Wu C. Million trees Los Angeles canopy cover and benefit assessment. Landscape and Urban Planning. 2011;**99**:40-50

[136] Asher M, Bhandari P. Mitigation or myth? Impacts of hydropower development and compensatory afforestation on forest ecosystems in the high Himalayas. Land Use Policy. 2021;**100**:105041

Chapter 2

# Forest Carbon Stock and Management Strategies for Riverine Forests in Sudan

*Abubakr A.A. Osman and Emad H.E. Yasin*

## Abstract

Riverine forests play a crucial role in carbon sequestration and sustainable forest management in Sudan. This chapter explores the species' ecological significance, growth patterns, and economic importance, while highlighting its contribution to mitigating climate change through carbon stock accumulation. Riverine forests are vital components of Sudan's ecosystem, supporting biodiversity, local livelihoods, and the timber industry. The chapter further examines forest management strategies, including thinning, volume estimation, and carbon stock assessment methods, which are essential for sustainable utilization. The importance of accurate biomass estimation, both destructive and non-destructive, is underscored to ensure effective carbon stock monitoring. Given the increasing global concerns over greenhouse gas emissions, the role of Sudan's riverine forests in carbon sequestration is critical. Sustainable management strategies, including afforestation, controlled harvesting, and conservation policies, are necessary to maintain these forests' ecological and economic benefits. Through proper implementation of these strategies, riverine forests can contribute to climate change mitigation and sustainable land-use practices in Sudan.

**Keywords:** riverine forests, carbon sequestration, forest management, biomass estimation, climate change mitigation, sustainable forestry

## 1. Introduction

Riverine forests play a crucial role in the global carbon cycle through their dynamic exchange of $CO_2$ with the atmosphere [1–4]. These unique ecosystems serve as vital carbon sinks, absorbing substantial amounts of atmospheric $CO_2$ during photosynthesis and storing it as fixed biomass in trees, shrubs, and soil organic matter [5–8]. This carbon sequestration function is particularly important in Sudan, where riverine forests along the Nile and other waterways form ecological corridors that maintain regional climate stability and support biodiversity. However, these critical ecosystems face mounting threats from agricultural expansion, unsustainable logging, and land-use changes that are altering their biomass and carbon-storage capacity [9, 10]. The carbon stocks in Sudan's riverine forests are particularly vulnerable to fluctuations caused by selective harvesting, regeneration patterns, and conversion to farmland, all



IntechOpen

of which significantly impact carbon fluxes to the atmosphere and local temperature regulation [11, 12]. These pressures create an urgent need for accurate assessment and sustainable management of these forest resources to maintain their climate-regulation services.

The pressing need for precise measurement of carbon storage in Sudan's riverine forests reflects broader global concerns about rising atmospheric $CO_2$ levels [9, 13, 14]. As carbon emissions continue to increase worldwide, international agreements like the Kyoto Protocol have established frameworks for emission reductions, with particular attention to forest carbon stocks in developing nations [15–18]. Sudan's participation in such agreements creates obligations to monitor and report on forest carbon stocks, making accurate measurement essential for verifying progress toward climate commitments [16, 17, 19]. The United Nations Framework Convention on Climate Change (UNFCCC) has further emphasized the critical importance of reliable carbon-stock assessments for all nations, particularly those with significant forest resources like Sudan [20]. Beyond compliance with international agreements, robust carbon measurement systems are fundamental for developing effective climate change mitigation strategies at national and local levels [20–23]. In the Sudanese context, this means creating tailored approaches that account for the unique characteristics of riverine forests, including their linear distribution, species composition, and hydrological dependencies.

The management of Sudan's riverine forest carbon stocks represents a significant opportunity for climate change mitigation while addressing local development needs [2, 4, 10, 24]. These ecosystems are particularly efficient carbon sinks due to their high productivity and substantial biomass accumulation in relatively narrow riparian zones [1, 14, 25, 26]. As these forests grow, they continuously remove carbon from the atmosphere, storing it in woody tissues, leaves, and root systems [6, 9, 12, 14]. However, realizing their full carbon sequestration potential requires implementation of science-based management strategies that balance conservation with sustainable use [4, 27]. Key approaches include selective thinning to enhance the growth of high-carbon species, precise volume estimation to guide harvesting decisions, and regular carbon stock assessments to monitor changes over time [28]. The accuracy of these management interventions depends fundamentally on reliable biomass estimation methods, whether through destructive sampling of representative trees or non-destructive techniques using allometric equations and remote sensing technologies. Such methods must be carefully calibrated for Sudan's unique riverine forest species, which may differ significantly from the global models typically used in carbon accounting [29–32].

The challenges facing Sudan's riverine forests highlight the complex interplay between climate change mitigation, ecosystem conservation, and human development needs [33]. On one hand, these forests represent a natural climate solution that can contribute to global carbon sequestration efforts while providing essential ecosystem services like water filtration, flood control, and habitat provision [34, 35]. On the other hand, they face intense pressure from local communities who rely on them for fuelwood, construction materials, and agricultural land [36]. This tension underscores the need for management strategies that recognize both the global importance of forest carbon stocks and the local realities of resource dependence [4, 37]. Recent advances in carbon measurement technologies, including improved allometric equations for African tree species and the application of satellite-based monitoring systems, offer new opportunities for tracking carbon stocks in Sudan's riverine forests with greater precision and lower costs [29, 30, 38]. These technological developments, combined with community-based forest management approaches, could help reconcile conservation goals with local livelihoods while contributing to international climate commitments [30]. The case of Sudan's riverine forests thus presents both a significant challenge and a promising opportunity for demonstrating how tropical forest ecosystems can be effectively managed for carbon sequestration alongside other ecological and social benefits [39]. As climate change impacts intensify across Africa, the preservation and sustainable management of these critical ecosystems will become increasingly important for both climate adaptation and mitigation efforts in the region [40, 41]. Therefore, this chapter further examines forest management strategies, including thinning, volume estimation, and carbon stock assessment methods, which are essential for sustainable utilization. The importance of accurate biomass estimation, both destructive and non-destructive, is underscored to ensure effective carbon stock monitoring.

## 2. Methodology

We relied on secondary data sources to accomplish the objectives outlined in the book chapter. Specifically, we gathered credible and relevant information using Scopus, Google Scholar, and Web of Science. These platforms were chosen because they offer extensive access to high-quality scientific publications and are equipped with user-friendly search features. Additionally, their widespread use in academic institutions ensures availability through university subscriptions.

## 3. Results and discussion

### 3.1 Forest resources and *Acacia nilotica* in Sudan

#### *3.1.1 Forests resources in Sudan*

Forests are indispensable at the local, national, and global levels, providing critical resources and environmental stability. Local communities, particularly those near forests, depend on them for food, fuel, and raw materials [42, 43]. However, excessive and unregulated exploitation has degraded many forests, altering their ecological composition [44, 45].

Nationally, forests contribute significantly to economies through timber, lumber, and other wood-based industries [24, 46]. Despite the rise of alternative materials, global timber demand continues to grow, increasing pressure on forest ecosystems. Beyond economic benefits, forests play a fundamental role in sustaining environmental systems by regulating climate, preserving biodiversity, and maintaining water cycles [47, 48].

Forests offer both tangible and intangible benefits. They supply timber and non-timber products, while delivering essential ecological services such as flood control, carbon sequestration, and soil conservation [10, 49]. Although large, productive forests still exist, their finite nature demands sustainable management to prevent irreversible depletion [26, 50].

The loss of forests threatens not only biodiversity but also global climate stability. Protecting these ecosystems is crucial for future generations, requiring balanced policies that support conservation alongside responsible resource use. Sustainable forestry practices, reforestation efforts, and stricter regulations are necessary to ensure forests continue supporting life on Earth.

### *3.1.2 Acacia nilotica species*

*Acacia nilotica (L.) Willd. ex Del.*, locally referred to as Sunt, is a valuable, multipurpose legume tree species. This plant has a broad distribution across the tropical and subtropical regions of Africa. The species comprises nine recognized subspecies, each occupying distinct geographical ranges. Notably, seven of these subspecies are native to various African regions, while four are specifically found in Sudan [51, 52].

#### *3.1.2.1 Description*

*Acacia nilotica* (family Leguminosae, subfamily Mimosoideae) typically reaches a height of 15–30 meters and a trunk diameter of 2–3 meters. In young trees, the bark is pale green and salty to the taste, while in mature specimens, it turns nearly black, with deep vertical fissures revealing a gray to pinkish inner layer that secretes a reddish, low-grade gum. The leaves are bipinnate, with 3–10 pairs of pinnae measuring 1.3–3.8 cm, each bearing 10–20 pairs of small leaflets approximately 2–5 mm long. Young trees are armed with slender, straight, light-gray spines arranged in axillary pairs, usually 3–12 in number and 5–7.5 cm long, though these are generally absent in mature trees. The flowers are bright golden yellow and form spherical heads, 1.2–1.5 cm in diameter, borne singly or in whorls on peduncles 2–3 cm long at the branch tips. The pods, measuring 7–15 cm, are green and fuzzy (tomentose) when immature, turning greenish-black and remaining indehiscent at maturity. They are deeply constricted between seeds, giving a bead-like appearance. Each pod typically contains 8–12 compressed, dark brown, glossy seeds with a hard outer coat [53–55].

#### *3.1.2.2 Growth pattern*

*Acacia nilotica* typically germinates following rainfall during the wet season. While the majority of seeds (around 95%) lose viability within 2 years, a small proportion can remain viable and germinate for up to 15 years after dispersal. Germination is enhanced by environmental disturbances, such as fire or passage through the digestive tract of animals. Seedling growth is vigorous in moist areas near water sources but significantly slower in open grassland environments. The species can begin flowering and fruiting as early as 2 to 3 years after germination, with reproductive activity often accelerated during years of abundant rainfall. Flowering generally occurs between March and June, followed by pod development from July to December. Leaf shedding predominantly takes place during the dry season, from June to November, and seedpods are typically dispersed between October and January [53–56].

#### *3.1.2.3 Distribution*

*Acacia nilotica (L.) Willd. ex Del.,* commonly referred to as Sunt, is a valuable multipurpose leguminous tree widely distributed across tropical and subtropical regions of Africa. The species comprises nine subspecies, each occupying distinct geographical zones, with seven found in Africa and four occurring specifically in Sudan [51, 52, 55]. In Sudan, the subspecies tomentosa and nilotica are prevalent, particularly in low-lying floodplains, known as Mayas along the Blue Nile and its

DOI: http://dx.doi.org/10.5772/intechopen.1010986

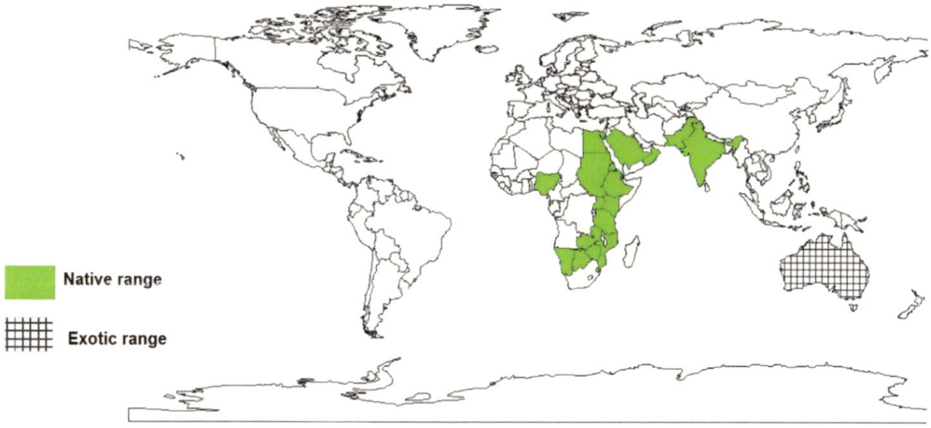


**Figure 1.**
*Map of documented native and exotic occurrences of the species.*

tributaries. These Mayas are seasonally inundated during floods, with water persisting for up to 6 months. Meanwhile, the subspecies adstringens is typically found along seasonal riverbanks and valleys in the Kordofan and Darfur regions (**Figure 1**) [55].

### *3.1.2.4 Ecology*

There is some evidence that *A. nilotica* is a weed in its native habitat, e.g., South Africa [53–57]. However, in many other regions, it is deliberately cultivated for forestry purposes and for rehabilitating degraded lands [57–59]. While *A. nilotica* is commonly used as a browse species, its overutilization—especially when appropriate stocking rates are not maintained—can contribute to land degradation. The species thrives particularly well in riverine alluvial soils and black cotton soils, and it is also capable of growing in saline, alkaline, and calcareous pan soils [60]. *A. nilotica* is adapted to a range of climates from subtropical to tropical and can tolerate extreme temperatures exceeding 50°C as well as drought conditions. Nonetheless, it requires sufficient moisture for optimal growth. Young plants are vulnerable to frost, and trees of all age classes can be negatively impacted by severe frost. The species is also sensitive to fire, with both seedlings and saplings being particularly susceptible. Its natural habitat spans regions with an average annual rainfall between 250 and 1500 mm [55, 61].

### *3.1.2.5 Thinning*

When *Acacia nilotica* stands become overcrowded, it is essential to thin them promptly to restore optimal growing conditions. Thinning practices, especially in riverain forests, should follow the principle of "thinning from below," which involves removing suppressed and intermediate trees to promote the growth of dominant individuals while maintaining a relatively closed canopy. Depending on site quality, commercial thinning can begin between the ages of 6 and 20 years [62–64].

Plantations of *A. nilotica* located between Sennar and Damazin are typically managed on a 30-year rotation cycle, primarily for the production of railway sleepers. To sustain this management system, plantations are subject to a systematic thinning schedule and yield regulation plan [65].

Jackson [63] proposed a five-stage thinning regime for *A. nilotica* plantations, recommending thinning at ages 6, 9, 12, 15, and 20 years. By the 20th year, the stand should be reduced to approximately 50 trees per feddan (equivalent to 125 trees per hectare), with a spacing of roughly 9 × 9 meters [65–67].

## 3.2 Forest carbon stock and measurement techniques

### *3.2.1 Volume estimation*

Tree volume is commonly estimated using direct volume equations, particularly for individual trees or tree groups. In this method, the tree is modeled as a geometric solid, most often a cylinder. The basic formula used is:

$$\mathbf{V} = \frac{\pi}{4} * D^2 * H \tag{1}$$

where $D$ is the diameter at breast height (DBH) and $H$ is the total height, which yields the volume of an idealized cylinder. However, since actual tree shapes deviate from perfect cylinders, a form factor is applied to adjust the calculated volume. This form factor serves as a reduction coefficient that accounts for taper and shape irregularities, thereby providing a more accurate estimate of true tree volume [68–70].

### *3.2.2 Volume table*

A volume table is a tabulated tool that provides average tree volume estimates based on one or more measurable tree dimensions. It expresses the relationship between tree volume (the dependent variable) and readily measurable attributes such as diameter at breast height (DBH) and total height (independent variables) [71].

The primary goal of constructing a volume table is to enable fast, cost-effective, and non-destructive estimation of tree volume using easily obtainable field measurements. Once an accurate volume has been determined for a tree, it can be applied to other trees of similar diameter, height, form, and growing conditions within the same locality [70, 72].

### *3.2.3 Types of volume table*

Volume tables can be categorized into several types based on their construction approach and the number of independent variables they incorporate. These include local volume tables, which are specific to species or a particular site or region; general volume tables, applicable across broader areas; form-class volume tables, which adjust for tree form variations; harmonized curve volume tables, based on smoothed volume curves; and equation-based volume tables, which rely on mathematical volume functions [73, 74].

### *3.2.4 Tariff system*

The term *tariff*, of Arabic origin, has long been used in European forestry to refer to specific types of local volume tables that estimate tree volume based solely on diameter measurements [68, 73–75]. Tariff tables are typically constructed using the linear relationship between tree volume and basal area, particularly within even-aged forest stands [68, 73–75]. Two tariff systems have been developed: the odd-numbered tariff system (**Table 1**) by Hetherington and Elsiddig [73] and the even-numbered tariff system (**Table 2**) by Osman [76].

### *3.2.5 Greenhouse gas overviews*

Greenhouse gases (GHGs) are components of the atmosphere that trap heat, thereby contributing to the greenhouse effect. These gases include both naturally occurring substances, such as carbon dioxide ($CO_2$), which are released through natural processes and human activities, and synthetic gases like fluorinated compounds that are exclusively anthropogenic [23]. Carbon dioxide is the most significant greenhouse gas emitted through human activities. In 2014, $CO_2$ made up approximately 80.9% of greenhouse gas emissions in the United States attributed to human sources [77]. Although $CO_2$ is a natural part of the Earth's carbon cycle—circulating among the atmosphere, oceans, terrestrial systems, and living organisms—human activities have disrupted this balance by both increasing emissions and impairing the function of natural carbon sinks such as forests. The rise in atmospheric $CO_2$ concentrations since the Industrial Revolution is primarily due to human-driven emissions, despite the continued presence of natural sources **Figure 2** [79–81].

| Tariff no. | Equation | R2 |
|---|---|---|
| 3 | V = 3.599 g - 0.009 | 0.889 |
| 5 | V = 5.343 g - 0.024 | 0.960 |
| 7 | V = 7.545 g - 0.024 | 0.986 |
| 9 | V = 9.640 g - 0.024 | 0.998 |
| 11 | V = 11.541 g - 0.040 | 0.996 |
| 13 | V = 13.499 g - 0.047 | 0.989 |
| 15 | V = 15.433 g - 0.056 | 0.990 |
| 17 | V = 17.13 g - 0.066 | 0.990 |
| 19 | V = 19.18 g - 0.060 | 0.989 |

**Table 1.**
*Tariff equations for odd-numbers [73].*

| Tariff no. | Equation | R2 |
|---|---|---|
| 4 | V = 0.006 + 4.407 g | 0.965 |
| 6 | V = 0.009 + 6.256 g | 0.982 |
| 8 | V = 0.02 + 7.963 g | 0.980 |
| 10 | V = 0.002 + 9.809 g | 0.989 |
| 12 | V = 0.025 + 12.118 g | 0.963 |
| 14 | V = 0.167 + 14.654 g | 0.988 |

**Table 2.**
*Tariff equations for even-numbers [76].*

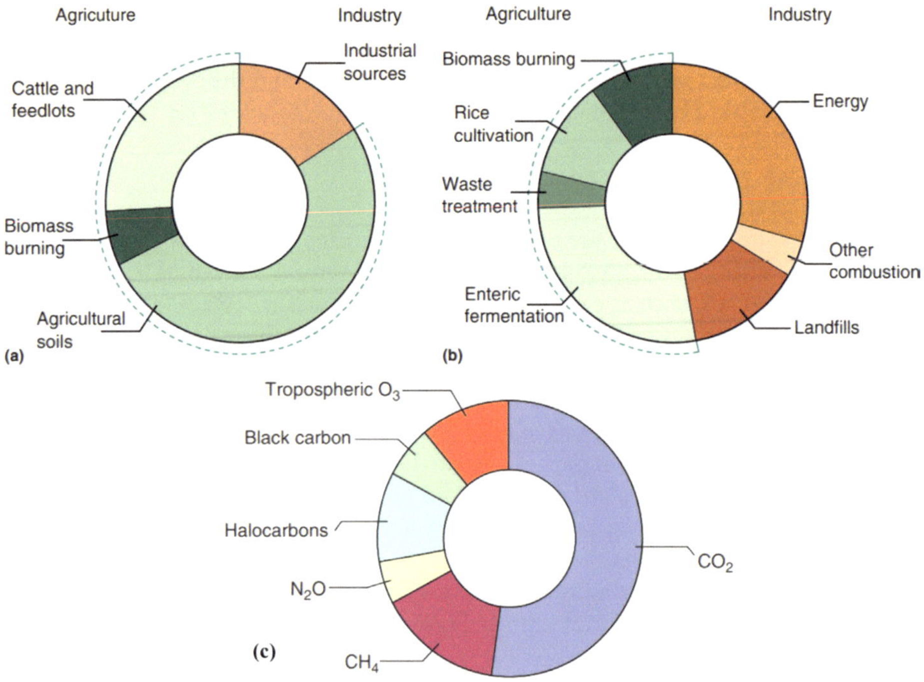


**Figure 2.**
*Anthropogenic sources of (a) nitrous oxide and (b) methane [78].*

### *3.2.6 Carbon pools and carbon stock*

Carbon stock refers to the quantity of carbon stored within a given "pool," which is a reservoir or system capable of accumulating or releasing carbon over time [82, 83]. These pools can store carbon through both physicochemical and biological processes. Depending on the scale and context, a carbon pool may represent a vegetation type, an ecosystem, a specific land area, or even an entire country [84, 85]. In terrestrial ecosystems, five principal carbon pools are typically recognized: above-ground biomass, below-ground biomass, dead wood, litter, and soil organic matter [9, 38, 86]. Carbon dioxide ($CO_2$), fixed by plants during photosynthesis, is allocated across these pools, with above-ground biomass often forming the largest and most visible component of carbon storage in forest ecosystems [61, 87, 88].

According to the GRID-Arendal/UNEP [89], the largest carbon stocks are generally found in tropical regions and high-latitude ecosystems. In tropical areas, carbon is predominantly stored in vegetation biomass, while in high-latitude ecosystems, it is mostly sequestered in permafrost and peat soils. The amount of carbon stored in an ecosystem varies by species composition, climate, soil type, and other ecological variables. Terrestrial ecosystems overall store about three times more carbon than the atmosphere, with an estimated 2100 gigatonnes (Gt) of carbon sequestered in vegetation, litter, and soil organic matter [9].

Within forest ecosystems, approximately 46% of total carbon stocks are found, with 39% located in forest soils, including their organic layers. However, the distribution of carbon varies significantly between tropical and boreal forests. For instance, while tropical forests store about 50% of their carbon in soils, boreal forests store approximately 84% in soil layers. Notably, the boreal forests of Russia, Canada, and

Alaska alone are responsible for nearly half of the global forest carbon stock, whereas tropical forests contribute around 37% [9, 14, 85, 90].

#### *3.2.7 Estimation biomass and carbon*

Estimating the accumulated biomass within forest ecosystems is crucial for evaluating forest productivity and sustainability. It provides insights into the potential carbon emissions—particularly in the form of carbon dioxide ($CO_2$)—that may be released during deforestation or forest-burning events [14, 91]. In addition, biomass estimation allows for the calculation of the amount of atmospheric $CO_2$ that forests are capable of sequestering, thereby contributing to climate change mitigation efforts.

Accurate assessments of forest biomass are essential for a range of applications, including sustainable timber extraction, monitoring carbon stock changes, and analyzing the dynamics of the global carbon cycle. Forest biomass can be measured using field-based methods, such as destructive sampling and allometric equations, or through remote sensing and GIS technologies, which offer broader spatial coverage and repeated temporal monitoring [1, 14, 30, 88, 92].

##### *3.2.7.1 Field-based biomass estimation methods*

There are two primary field methods used for estimating forest biomass. The first is the destructive method, often referred to as the harvest method, which is considered the most direct and accurate approach for determining above-ground biomass and carbon stock in forest ecosystems [9, 13, 38, 91, 93]. This technique involves felling all trees within a defined plot and separately weighing the components—such as stems, branches, and foliage—both in fresh and oven-dried conditions to determine dry biomass [75, 94–97].

Despite its accuracy, the destructive method is limited in scope due to its labor-intensive, time-consuming, and costly nature. It is also unsuitable for large-scale applications and cannot be used in degraded or conservation-sensitive areas with rare or endangered species [32, 91, 98, 99]. However, it remains valuable for developing allometric equations used to estimate biomass over larger regions (**Figure 3**) [5, 101, 102].

The second method is non-destructive biomass estimation, which allows for biomass assessment without harvesting trees. This approach is especially relevant for ecosystems containing protected or threatened species, where felling is not feasible [91, 98, 99]. Non-destructive techniques typically involve climbing trees to measure physical parameters or using measurements such as diameter at breast height (DBH), tree height, volume, and wood density to estimate biomass through allometric equations [91, 102, 103].

### 3.3 Sustainable management strategies for *Acacia nilotica*

*Acacia nilotica* requires integrated management approaches to balance ecological benefits with socioeconomic value. Silvicultural practices must optimize productivity while maintaining sustainability. Studies show that 5–10 year rotation cycles can yield 10–25 $m^3$ $ha^{-1}$ y with improved genotypes and resource conservation, enhancing long-term productivity [104, 105]. However, intercropping risks in arid zones necessitate careful planning as it can significantly reduce crop yields [106, 107].

Community-based forest management (CBFM) has proven effective for *A. nilotica* conservation. Successful cases in Sudan and Uganda demonstrate that empowering

**Figure 3.**
*Destructive sampling process for tree biomass estimation: A = Stands were cut at a height of 40 cm diagonally, B = Fresh biomass measuring of biomass components, C = Biomass sampling, and D = Biomass drying [100].*

local communities enhances forest management [108, 109], while in Tanzania, CBFM has led to effective woodland protection and flora recovery [110, 111]. Nepal's Annapurna Conservation Area shows higher forest basal area and biodiversity in community-managed areas [112]. Key success factors include community participation, clear resource boundaries, and tangible benefits for local people.

The species' climate resilience makes it valuable for adaptation strategies. *A. nilotica* can withstand extreme temperatures and drought, making it suitable for arid land rehabilitation [53, 113]. Its potential distribution in Australia is expected to expand due to increased water-use efficiency from rising $CO_2$ levels [114, 115]. Biotechnological advances like direct regeneration from nodal segments and indirect organogenesis support propagation, though their invasive potential requires careful management [116, 117].

Policy frameworks must align ecological and economic goals. Initiatives like REDD+ can build on existing land management strategies [118], while successful Sahel restoration practices include enclosures and assisted regeneration [119]. Effective reforestation should incorporate ecological indicators and community engagement [120], with secure land tenure and benefit-sharing ensuring sustainability [118]. These integrated approaches can maximize *A. nilotica*'s contributions to sustainable development while addressing management challenges.

### 3.4 Carbon sequestration potential and deforestation challenges in Sudan

Sudan faces significant environmental challenges due to deforestation and forest degradation, primarily driven by agricultural expansion, unsustainable wood harvesting, and fuelwood consumption for brick production. These activities have led to substantial carbon emissions and reduced sequestration capacity. For instance, the brick-making industry alone consumes 508.4 × $10^3$ $m^3$ of wood biomass annually, emitting 455,666 t $CO_2$-equivalent [121]. Land-use changes have further altered carbon dynamics, with dense forest cover declining while bare land and light forest cover increase [122, 123]. In Ghana's Sudan savanna zone, forest degradation released 554,684.96 Mg $CO_2$, with only 31.84% offset by sequestration [124].

Despite these pressures, certain tree species, particularly *Acacia nilotica* and related varieties, show strong potential for carbon sequestration. In Sudan, *Acacia seyal* outperformed *Eucalyptus microtheca*, storing 370 tons of $CO_2$ per hectare compared to 176 tons [125]. *A. nilotica* also demonstrates high sequestration capacity, with an estimated 11.92 tons/hectare in Indonesian stands [126, 127] and the second-highest soil nitrogen content among 15 tree species, indicating robust carbon storage potential. However, as an invasive species, its management requires careful planning to avoid ecological disruption.

National and subnational forest conservation policies have had limited success in mitigating deforestation, particularly in tropical regions [128]. While India has projected increased forest cover through policy interventions, Sudan's efforts face implementation gaps. Community-based forest management has shown promise in improving sustainability [129], but stronger enforcement and local engagement are needed. Sustainable practices like agroforestry and native species cultivation can enhance livelihoods while aiding carbon mitigation [39].

Renewable energy development offers another pathway to reduce emissions. Sudan has abundant solar, wind, and biomass resources [130, 131], with *A. nilotica* contributing to biomass production. Expanding renewables could curb deforestation linked to fuelwood demand while improving energy security.

Overall, Sudan's deforestation crisis demands urgent action. Strengthening forest conservation, promoting high-sequestration species like *Acacia*, and investing in renewable energy are essential to restoring carbon balances and achieving climate resilience. Without effective policies and sustainable land-use practices, emissions will continue to rise, exacerbating environmental degradation.

### 3.5 Economic importance and utilization of *Acacia nilotica*

*Acacia nilotica* is a highly valuable multipurpose tree species with significant economic importance across Africa and the Indian subcontinent. It has been traditionally utilized for tannin production, timber, fuelwood, fodder, and medicinal applications.

#### *3.5.1 Industrial and commercial uses*

The bark and seeds of *A. nilotica* are rich in tannins, making them essential for leather tanning [132]. Its gum, though inferior to gum arabic (*A. senegal*), serves as a substitute in food and pharmaceutical industries [133]. The wood is exceptionally durable, shock-resistant, and nearly twice as hard as teak, making it ideal for construction, tool handles, and cart manufacturing. Additionally, its high calorific value (4950 kcal/kg) and slow-burning properties make it a preferred fuelwood and charcoal source.

#### *3.5.2 Medicinal applications*

*A. nilotica* has been widely used in traditional medicine. The bark treats hemorrhages, colds, diarrhea, tuberculosis, and leprosy, while the roots are used as an aphrodisiac, and the flowers help heal syphilis lesions [134]. It also acts as a powerful astringent, molluscicide, and algaecide.

#### *3.5.3 Agroforestry and land rehabilitation*

Subspecies indica and cupressiformis are commonly planted as windbreaks and for soil stabilization. In India, *A. nilotica* thrives in degraded saline-alkaline soils (pH up to 9) and is used to rehabilitate wastelands, including coal mine spoils and tannery effluent-affected areas [135]. Over 50,000 hectares of the Chambal ravines have been successfully restored using this species.

Given its diverse economic uses—ranging from timber and fuel to medicine and land reclamation—*A. nilotica* remains a crucial species for sustainable livelihoods and environmental management [136].

### *3.6 Challenges and future perspectives in Acacia nilotica management*

Accurate carbon stock measurement in *Acacia nilotica* plantations faces significant technical barriers. Tree height measurement proves more challenging than diameter at breast height assessments, impacting biomass calculations [137]. Stand density variations affect biomass distribution, while soil nitrogen storage fluctuates with stand age and depth [138, 139]. Current, allometric models often produce inaccurate estimates for mixed-species forests, necessitating species-specific equations [140]. These measurement challenges highlight the need for improved protocols considering stand age, density, and soil composition.

Emerging technologies offer promising solutions for monitoring *A. nilotica* ecosystems. Remote sensing tools like LiDAR and drone imagery provide detailed forest structure data [141], while AI algorithms enable vegetation classification and change detection [142]. However, developing countries face implementation barriers due to limited infrastructure [142], underscoring the need for capacity building.

Policy gaps significantly hinder conservation efforts. West Africa's experience with neglected crops demonstrates the importance of institutional capacity [143], while South Africa's wetland conservation struggles reveal policy harmonization challenges [144]. Effective governance requires multi-scalar approaches [145] and better implementation of international agreements [146].

Reforestation strategies must address ecological and social factors. *A. nilotica* shows tolerance to moderate salinity (4 g/l NaCl), suggesting potential for saline land rehabilitation [147]. Successful restoration requires integrated approaches combining soil management, native species use, and community engagement [148, 149]. Karst region studies emphasize the need for social-ecological frameworks [150], highlighting the importance of holistic planning for large-scale reforestation programs.

## 4. Conclusion

Riverine forests, particularly *Acacia nilotica*, play a pivotal role in Sudan's forest ecosystems, offering ecological, economic, and climate-related benefits. These forests serve as critical carbon sinks, significantly contributing to global and regional climate change mitigation efforts through carbon sequestration. However, unsustainable land-use practices, deforestation, and climate variability threaten their ability to provide long-term environmental services. Effective forest management strategies, including controlled thinning, volume estimation, biomass assessments, and carbon stock monitoring, are essential to maintain sustainability. Advances in remote sensing, GIS, and AI-driven carbon estimation models can further enhance the accuracy of forest carbon stock assessments, aiding in better conservation planning. Additionally, the economic significance of riverine forests extends to timber production, fuelwood supply, gum Arabic extraction, and medicinal applications, emphasizing the need for sustainable utilization. The integration of community-based forest management, agroforestry practices, and afforestation initiatives can promote conservation while ensuring local livelihoods. Addressing these challenges requires strong governance frameworks, adherence to international climate agreements (e.g., Kyoto Protocol, Paris Agreement, REDD+), and collaborative policy interventions. If managed effectively, Sudan's riverine forests can enhance biodiversity conservation, boost economic resilience, and contribute to global carbon reduction efforts, reinforcing their crucial role in climate change adaptation and mitigation.

## Author details

Abubakr A.A. Osman[1,2]* and Emad H.E. Yasin[1,2]

1 Department of Forest Management, Faculty of Forestry, University of Khartoum, Khartoum North, Sudan

2 Institute of Gematics and Civil Engineering, Faculty of Forestry, University of Sopron, Sopron, Hungary

*Address all correspondence to: bikoosman103@gmail.com

## References

[1] Chang Z, Xiong X, Zhang S, Wu J. Modeling and remote sensing of the Forest ecosystem. Forests. 2025;**16**(1):101

[2] Hurteau MD. The role of forests in the carbon cycle and in climate change. In: Climate Change. Amsterdam (Netherlands): Elsevier; 2021. pp. 561-579

[3] Munassar S, Rödenbeck C, Gałkowski M, Koch FT, Totsche KU, Botía S, et al. To what extent does the $CO_2$ diurnal cycle impact flux estimates derived from global and regional inversions? Atmospheric Chemistry and Physics. 2025;**25**(1):639-656

[4] Papa CC, Clay K, Cooper LT, Stark SC. Science-based communication and education needed to improve forest carbon science, policy, and management outcomes. Environmental Research Letters. 2025;**20**(2):024044

[5] Ioannidou S, Litskas VD, Stavrinides M, Vogiatzakis IN. The role of mixed orchards in carbon sequestration and climate change mitigation in a Mediterranean island environment. Frontiers in Sustainable Food Systems. 2025;**8**:1457462

[6] Matthews E, Payne R, Rohweder M, Murray S. Forest ecosystem: Carbon storage sequestration, carbon sequestration in soil. Global Climate Change Digest. 2000;**12**(2):19-99

[7] Sinyoung S, Jeeraro A, Udomkun P, Kunchariyakun K, Graham M, Kaewlom P. Enhancing $CO_2$ sequestration through corn stalk biochar-enhanced mortar: A synergistic approach with algal growth for carbon capture applications. Sustainability. 2025;**17**(1):342

[8] Victorero F, Bustamante W. Timber biogenic carbon stock in the urban environment: Santiago City as a second Forest. Sustainability. 2025;**17**(2):529. DOI: 10.3390/su17020529

[9] Duque A, Gómez-Correa LF, Barona-Colmenares AA, Castaño N, Davies S, Zuleta D, et al. Local scale carbon stock measurements, including deep soil layers, in a terra firme forest in northwestern Amazon. Journal of Tropical Ecology. 2025;**41**:e2

[10] Li C, Wu J, Zhang F, Huang X. Forest carbon sinks in Chinese provinces and their impact on sustainable development goals. Forests. 2025;**16**(1):83

[11] Segura R, Lauvaux T, Lian J, Ciais P, Badia A, Ventura S, et al. Heat and Drought Events Alter Biogenic Capacity to Balance $CO_2$ Budget in South-Western Europe. Washington, DC, USA: Authorea Preprints; 2024. e-location ID (e2024GB008163)

[12] Sipra M, Gobinda AC, Subash SC, Rupa TR, Dibya KS, Abhiram D, et al. Changes in carbon sequestration with age of trees in guava orchards at different locations in a tropical climate on Alfisols. Plant Science Today. 2025;**12**(1). DOI: 10.14719/pst.4456

[13] Ma Y, Zhang X, Jiang R, Jiang M, Ju J. Effects of stand structural characteristics, diversity, and stability on carbon storage across different densities in natural forests: A case study in the Xiaolong Mountains, China. Forests. 2025;**16**(1):71

[14] Zhou Y, Wei G, Wang Y, Wang B, Quan Y, Wu Z, et al. Estimating regional forest carbon density using remote sensing and geographically weighted

random forest models: A case study of mid-to high-latitude forests in China. Forests. 2025;**16**(1):96

[15] Abiodun TP, Nwulu N, Olukanmi P. Application of blockchain technology in carbon trading market: A systematic review. IEEE Access. 2024;**12**:e2024.3523672

[16] Khosravi H, Raihan AS, Islam F, Nimbarte A, Ahmed I. A comprehensive approach to $CO_2$ emissions analysis in high-human-development-index countries using statistical and time series approaches. Sustainability. 2025;**17**(2):603. DOI: 10.3390/su17020603

[17] Mravcová A. Assessing the effectiveness of international climate agreements in mitigating global warming. Studia Ecologiae et Bioethicae. 2025;**23**(1):51-62

[18] Ravindranath NH, Somashekhar BS, Gadgil M. Carbon flow in Indian forests. Climatic Change. 1997;**35**(3):297-320. DOI: 10.1023/A:1005303405404

[19] den Elzen MG, Dafnomilis I, Nascimento L, Beusen A, Forsell N, Gubbels J, et al. Uncertainties around net-zero climate targets have major impact on greenhouse gas emissions projections. Annals of the New York Academy of Sciences. 2025;**35**:297-320

[20] Yang H, Song M, Son H, Kim R, Choi E. Evaluating REDD+ readiness: High-potential countries based on MRV capacity. Forests. 2025;**16**(1):67

[21] Allan RP, Arias PA, Berger S, Canadell JG, Cassou C, Chen D, et al. Intergovernmental panel on climate change (IPCC). Summary for policymakers. In: Climate Change 2021: The Physical Science Basis. Contribution of Working Group I to the Sixth Assessment Report of the Intergovernmental Panel on Climate Change. Cambridge (UK) and New York (USA): Cambridge University Press; 2023. pp. 3-32

[22] Onat NC, Mandouri J, Kucukvar M, Kutty AA, Al-Muftah AA. Driving sustainable business practices with carbon accounting and reporting: A holistic framework and empirical analysis. Corporate Social Responsibility and Environmental Management. 2025;**32**:2795-2814

[23] Prakash O. Addressing the climate crisis: The urgent need for sustainable resource management. Journal of Environmental Biology. 2025;**46**(1):3-4

[24] Gao L, Li H, Li S. Effects on carbon sequestration of biomass and Investment in state-owned forest farms: A case study of Shaanxi Province, China. Forests. 2025;**16**(1):60

[25] Schlesinger WH. Biogeochemistry: An Analysis of Global Change. 2nd ed. New York: Academic Press; 1997

[26] Wright SJ, Calderón O. Statistical power and the detection of global change responses: The case of leaf production in old-growth forests. Ecology. 2025;**106**(1):e4526

[27] Takeda N, Rowlings D, Parton W, Grace L, Day K, Nguyen T, et al. Soil carbon sequestration potential in subtropical grasslands estimated by DayCent-CABBI. Soil Science Society of America Journal. 2025;**89**(1):e70003

[28] Boukhris I, Collalti A, Lahssini S, Dalmonech D, Nakhle F, Testolin R, et al. Timber Tracer: A comprehensive framework for the evaluation of carbon sequestration by forest management and substitution of harvested

wood products. bioRxiv. 2024. DOI: 10.1101/2024.01.24.576985

[29] Agboka KMJ, Soro D, Agboka K. Carbon sequestration potential of cashew (*Anacardium occidentale* L.) plantations across climatic gradients in Togo. International Journal of Environment and Climate Change. 2025;**15**(1):39-54

[30] Ali G, Mijwil MM, Adamopoulos I, Ayad J. Leveraging the Internet of Things, remote sensing, and artificial intelligence for sustainable forest management. Babylonian Journal of Internet of Things. 2025;**2025**:1-65

[31] Lin C, Doyog ND. Applying a four-way factorial experimental model to diagnose optimum kNN parameters for precise aboveground biomass mapping. IEEE Journal of Selected Topics in Applied Earth Observations and Remote Sensing. 2024;**17**:12345-12358

[32] Nkoua Ngavouka MD, Mayala TS, Douma DH, Brown AE, Hammerton JM, Ross AB, et al. Characterization of Congolese Woody biomass and its potential as a bioenergy source. Applied Sciences. 2025;**15**(1):371

[33] Mumtaz M, Jahanzaib SH, Hussain W, Khan S, Youssef YM, Qaysi S, et al. Synergy of remote sensing and geospatial technologies to advance sustainable development goals for future coastal urbanization and environmental challenges in a riverine megacity. ISPRS International Journal of Geo-Information. 2025;**14**(1):30

[34] Gao X. Climate change and the poor: Reasons, impacts, inequities, and sustainable solutions. Applied and Computational Engineering. 2025;**124**:237-245

[35] Haubrock PJ, Stubbington R, Fohrer N, Hollert H, Jähnig SC, Merz B, et al. A holistic catchment-scale framework to guide flood and drought mitigation towards improved biodiversity conservation and human wellbeing. Wiley Interdisciplinary Reviews: Water. 2025;**12**(1):e70001

[36] Chongtham IR, Shahi S, Chikkanjegowda M. The impact of shifting from subsistence to cash crops on the livelihoods of the Soliga tribe in India. Current Agriculture Research Journal. 2024;**12**(3):1106-1115. DOI: 10.12944/CARJ.12.3.06

[37] Rodrigues AS. Accounting for functionality in the identification of global conservation priorities: Promises and pitfalls. Philosophical Transactions B. 2025;**380**(1917):20230209

[38] Fahmi MY, Muttaqin AD, Ardiansyah RE, Anggreini R, Luna Widyawati SS. Potential carbon content by mangrove forest along the coastline of trenggalek regency. Ecological Engineering & Environmental Technology (EEET). 2025;**26**(1):41-57. DOI: 10.12912/27197050/195343

[39] Musa FI, Sahoo UK. Role of sustainable forest management in poverty reduction and livelihood improvement in Sudan. A review. International Journal of Ecology and Environmental Sciences. 2023;**49**:449-456

[40] Kapuka A, Hlásny T. Climate change impacts on ecosystems and adaptation options in nine countries in southern Africa: What do we know? Ecosphere. 2021;**12**(12):e03860

[41] Leal Filho W, Azeiteiro UM, Balogun AL, Setti AFF, Mucova SA, Ayal D, et al. The influence of ecosystems services depletion to climate change adaptation efforts in Africa. Science of the Total Environment. 2021;**779**:146414

[42] García-Jácome SP, Jankovský M, Hoeben AD, Lindner M, Uzquiano S, Stern T, et al. Forest value chain resilience from a local perspective in five European countries: Analysis of predictors and co-drivers. Frontiers in Forests and Global Change. 2025;**7**:1461932

[43] Rexhepi B, Abdija X, Bajrami A, Iseni G. Ecological, Socio-Cultural, and Economic Importance of Non-Timber Forest Products in Shar Mountain (North Macedonia). Tirana, Albania: Richtmann Publishing; 2025

[44] Benítez Á, Nagua R, Medina J, Lapo G, Yangua-Solano E, Andrade-Hidalgo R. Bryophytes as indicators of disturbance in one of the last remnants of the mountain forests of El Oro Province, Ecuador. Plants. 2025;**14**(2):184

[45] Marsh CJ, Turner EC, Blonder BW, Bongalov B, Both S, Cruz RS, et al. Tropical forest clearance impacts biodiversity and function, whereas logging changes structure. Science. 2025;**387**(6730):171-175

[46] Zhang X, Han S, Zheng X, Chen Y. Does the regional comprehensive economic partnership promote forest product exports? Evidence on bilateral export performance from China. Forests. 2025;**16**(1):64

[47] FOSA. Sudan country outlook/ FNC-FAO. In: Forest Outlook Study for Africa Heatherington and Elsiddig 1983/1984. Basel, Switzerland: MDPI (Multidisciplinary Digital Publishing Institute); 2001

[48] Kallio AMI, Strîmbu VF, Gobakken H, Gobakken T. Climate change mitigation through alternative uses of Norwegian forest resources. Silva Fennica. 2025;**59**(1):23066. DOI: 10.14214/sf.23066

[49] Ren B, Xiao Y, Liu B, Geng J, Wu W, Qin D. Exploring the transmission process of carbon sequestration services and its applications: A case study of Hainan. Forests. 2025;**16**(1):136. DOI: 10.3390/f1601013617

[50] Ali OB. Assessment of changes on land cover and species composition and their impacts at Adar Yale, and Paloich Area, Upper Nile, [PhD thesis]. Shambat (Sudan): Faculty of Forestry, University of Khartoum; 2008. 130p

[51] Abuelmaali SA, Jamaluddin JAF, Noaman K, Allam M, Abushama HM, Elnaiem DE, et al. Distribution and genetic diversity of *Aedes aegypti* subspecies across the Sahelian Belt in Sudan. Pathogens. 2021;**10**(1):78

[52] Omer A, Kordofani M, Gibreel HH, Pyšek P, van Kleunen M. The alien flora of Sudan and South Sudan: Taxonomic and biogeographical composition. Biological Invasions. 2021;**23**(7):2033-2045

[53] Bargali K, Bargali SS. *Acacia nilotica*: A multipurpose leguminous plant. Nature and Science. 2009;**7**(4):11-19

[54] Raj A, Haokip V, Chandrawanshi S. *Acacia nilotica*: A multipurpose tree and source of Indian gum Arabic. South Indian Journal of Biological Sciences. 2015;**1**(2):66-69

[55] Rayan AA. Antibacterial influences of Accacia spp. in ed-duiem locality, White Nile province, Sudan. A chemical, pharmaceutical and traditional medicine overview. International Journal of Scientific World. 2025;**11**(1):10-12. DOI: 10.14419/9wzaa779

[56] Jame R. Phytochemical and pharmacological uses of *Acacia nilotica*—A review. Seeds. 2018;**1**:15-21

[57] Mwangi JG, Mohammed S, Umar KM, Haggar J, Santika T. Towards

the sustainability of African sandalwood: Understanding the distribution and environmental requirements. Plants, People, Planet. 2024;**6**:1-13

[58] Bobuľská L, Demková L, Pinčáková G, Lošák T. Alien plant invasion: Are they strictly Nature's enemy and how can we use their supremacy? Land. 2025;**14**(1):107

[59] Torbert JL, Burger JA. Forest land reclamation. Reclamation of Drastically Disturbed Lands. 2000;**41**:371-398

[60] Kachhadiya DD, Pillai P. A review on soil erosion and its control techniques. International Journal of Science and Research Archive. 2025;**14**(1):267-273

[61] Barras Rodriguez JD, Zuñiga Cespedes B, Castellanos Barliza JDJ. Evaluation of heat shock effects on seedling emergence and seed Bank viability in 18 Colombian dry forest species. Applied Vegetation Science. 2025;**28**(1):e70007

[62] Gorrod E, White L, Bedward M, Hope B, Chu N, McAllister D. Restoration thinning accelerates small tree growth but may slow large tree growth in a multi-age flood-dependent forest. Journal of Applied Ecology. 2024;**61**(7):1560-1572

[63] Jackson JK. Fung Working Plan For 1959-1968, for. Dep. (Sudan) for. Bull. No. 3. Khartoum, Sudan: Forests Department, Sudan; 1958. 71p

[64] Sheng H, Long F, Li X, Haider FU, Shi Z, Xian L, et al. Thinning improves large diameter timber cultivation but undermines ecosystem multifunctionality in the short term. Forests. 2025;**16**(1):134

[65] Amadou I, Soulé M, Salé A. An overview on the importance of *Acacia nilotica* (L.) willd. ex del.: A review. Asian Journal of Research in Agriculture and Forestry. 2020;**5**:12-18

[66] Elsiddig EA. The management of *Acacia nilotica*. L plantation along the Blue Nile flood basins in Sudan for sustainable production: A review. U. of K. The Journal of Agricultural Science. 2002;**10**(1):119-130

[67] Roy J, Akther HMR. Determination of financial rotation of *Acacia nilotica* (L.) plantations: A case study in Faridpur District of Bangladesh. Journal of Economics and Sustainable Development. 2021;**12**(4):34-40

[68] França LF. A model for roundwood of planted forests. Advances in Forestry Science. 2024;**11**(4):17558. DOI: 10.34062/afs.v11i4.17558

[69] Kershaw JA Jr, Ducey MJ, Beers TW, Husch B. Forest Mensuration. Chichester (UK) and Hoboken: John Wiley & Sons; 2016

[70] Netto SP, Stefanello FR, Nhaduco OPE, Madi JPS, Morales JJO, Behling A. Modeling tree form in forest stands of *Pinus taeda* L in southern of Brazil. Caderno Pedagógico. 2025;**22**(1):e13286-e13286

[71] Hegde R, Clara Manasa PA, Salimath SK. Forest mensuration. In: Textbook of Forest Science. Springer Nature Singapore: Singapore; 2025. pp. 361-388

[72] Cunia T. Weighted least squares method and construction of volume tables. Forest Science. 1964;**10**(2):180-191

[73] Hetherington JC, Elsiddig EA. Aids to the management of *Acacia nilotica* plantations in the Sudan. Forest Ecology and Management. 1984;7(4):275-289

[74] Rawat JK, Dasgupta S, Kumar R, Kumar A, Chauhan KVS. Training Manual on Inventory of Trees Outside Forests (TOF). Bangkok (Thailand): FAO; 2003. Available from: https://fao.org/4/ac840e/AC840E00.htm#TOC [Accessed: May 06, 2025]

[75] Utami WS, Herlambang H, Kiswanto K, Permadi A, Saud OR. Model Alometrik untuk menduga volume pohon berdiri Shorea leprosula Miq, PT Gunung Muro Wahana Jaya. Wahana Forestra: Jurnal Kehutanan. 2025;**20**(1):13-24

[76] Osman AAA. Estimation of biomass and carbon stock changes in Sunut plantation (Acacia Nilotica) in the Blue Nile, Sinnar State, Sudan [M.Sc thesis]. Sudan: University of Khartoum; 2018. pp. 1-70

[77] Wilson DD, Tefera GW, Ray RL. Application of Google earth engine to monitor greenhouse gases: A review. Data. 2025;**10**(1):8

[78] Robertson GP. Soil greenhouse gas emissions and their mitigation. Encyclopedia of Agriculture and Food Systems. 2014;**5**:185-196

[79] Lapola DM, Blanco CC, Cardeli BR, Esquivel-Muelbert A, Martinelli JV, Quesada CA, et al. Not just semantics: $CO_2$ fertilization can be a disturbance leading to worldwide forest degradation. Plants, People, Planet. 2024;7:638-643

[80] NRC. Advancing the Science of Climate Change of Semi-evergreen Tropical Forest of Manipur, North-Eastern India. Washington, DC, USA: National Research Council. The National Academies Press; 2010

[81] National Research Council, Division on Earth, Life Studies, Board on Atmospheric Sciences, America's Climate Choices, Panel on Advancing the Science of Climate Change. Advancing the Science of Climate Change. National Academies Press; 2011. pp. 1-501

[82] Keenan RJ, Reams GA, Achard F, de Freitas JV, Grainger A, Lindquist E. Dynamics of global forest area: Results from the FAO global forest resources assessment 2015. Forest Ecology and Management. 2015;**352**:9-20

[83] Suhardono S, Wahyuni T, Nazar IA, Ibriza NM, Wijayanti S. Green open space mapping and carbon stock estimation in Surakarta. E3S Web of Conferences. 2025;**605**:03047

[84] Wang W, Sheng L, Yu X, Zhang J, Su P, Zou Y. Soil–plant carbon pool variations subjected to agricultural drainage in Xingkai Lake wetlands. Water. 2025;**17**(1):125

[85] WGBU WBDBG, Beese F, Fraedrich K, Klemmer P, Kokott J, Kruse-Graumann L, et al. The accounting of biological sinks and sources under the Kyoto protocol: A step forward or backwards for global environmental protection? Special Report 1998. 1998

[86] Vashum KT, Jayakumar S. Methods to estimate above-ground biomass and carbon stock in natural forests-A review. Journal of Ecosystem & Ecography. 2012;**2**(4):1-7

[87] Qin K, Ye X, Luo S, Fernie AR, Zhang Y. Engineering carbon assimilation in plants. Journal of Integrative Plant Biology. 2025;**67**:926-948

[88] Ravindranath NH, Chaturvedi RK, Murthy IK. Forest conservation, afforestation and reforestation in India: Implications for forest carbon stocks. Current Science. 2008;**95**:216-222

[89] GRID-Arendal/UNEP. The Natural Fix? The Role of Ecosystems in Climate Mitigation. Nairobi: UNEP/WCMC; 2013

[90] Ķēniņa L, Bārdule A, Bičkovskis K, Desaine I, Jansons Ā. Preliminary assessment of stock of organic carbon in mineral soils of hemiboreal forests in Latvia. Baltic Forestry. 2024;**30**(2):id776-id776

[91] Pratama BA, Atikah TD, Susanti R, Jakalalana S, Ferdinandus SP, Maail RS. Stand structure and biomass estimation for cultivation of agarwood-producing species on Buru Island, Maluku Province, Indonesia. Media Konservasi. 2025;**30**(1):63-63

[92] Fachrun A, Rukmana D, Idrus R. Climate change control potential of mangrove ecosystems in Rawa Aopa Watumohai National Park, Tinanggea District. Ecological Engineering & Environmental Technology (EEET). 2025;**26**(1):379-387

[93] Gibbs HK, Brown S, Niles JO, Foley JA. Monitoring and estimating tropical forest carbon stocks: Making REDD a reality. Environmental Research Letters. 2007;**2**(4):045023

[94] Ceulemans R. Allometric relationships for below- and aboveground biomass of young scots pines. Forest Ecology and Management. 2004;**203**:177-186

[95] Pertiwi H, Matangaran JR. Quantification of logging residue from harvesting of Rasamala wood (*Altingia excelsa*). International Journal of Current Science Research and Review. 2025;**8**(1):26-32. DOI: 10.47191/ijcsrr/V8-i1-04

[96] Ravindranath NH, Ostwald M. Carbon inventory methods for National Greenhouse gas Inventory. In: Handbook for Greenhouse Gas Inventory, Carbon Mitigation and Roundwood Production Projects. Dordrecht, Netherlands: Springer; 2008. pp. 217-235

[97] Elsiddig EA, Hetherington JC. The Stem and Branch Volume of *Acacia nilotica* in the Fung Region in the Sudan. University College of North Wales, Department of Forestry and Wood Science; 1985

[98] Alvianingsih G, Hashim H, Jamian JJ, Senen A. Determination of biomass energy potential based on regional characteristics using adaptive clustering method. International Journal of Electrical & Computer Engineering. 2025;**15**(1):46-55. ISSN: 2088-8708. DOI: 10.11591/ijece.v15i1

[99] Montes N, Gauquelin T, Badri W, Bertaudiere V, Zaoui EH. A non-destructive method for estimating above-ground forest biomass in threatened woodlands. Forest Ecology and Management. 2000;**130**(1):37-46

[100] Tadese S, Soromessa T, Bekele T, Bereta A, Temesgen F. Above ground biomass estimation methods and challenges: A. Measurement. Journal of Energy Technologies and Policy. 2019;**9**(8):8-19

[101] Návar J. Allometric equations for tree species and carbon stocks for forests of northwestern Mexico. Forest Ecology and Management. 2009;**257**(2):427-434

[102] Qi Y, Chen X, Chen Z, Zhang X, Shen C, Chen Y, et al. Estimation model for cotton canopy structure parameters based on spectral vegetation index. Life. 2025;**15**(1):62

[103] Aboal JR, Arévalo JR, Fernández Á. Allometric relationships of different tree species and stand above ground biomass in the Gomera laurel forest (Canary Islands). Flora-Morphology, Distribution, Functional Ecology of Plants. 2005;**200**(3):264-274

[104] Huong VD, Nambiar ES, Hai NX, Ha KM, Dang NV. Sustainable

management of *Acacia auriculiformis* plantations for wood production over four successive rotations in South Vietnam. Forests. 2020;**11**(5):550

[105] Nambiar EKS, Harwood CE. Productivity of acacia and eucalypt plantations in Southeast Asia. 1. Bio-physical determinants of production: Opportunities and challenges. International Forestry Review. 2014;**16**(2):225-248

[106] Mihrete TB, Mihretu FB. Crop diversification for ensuring sustainable agriculture, risk management and food security. Global Challenges. 2025;**9**:2400267

[107] Puri S, Singh S, Kumar A. Growth and productivity of crops in association with an *Acacia nilotica* tree belt. Journal of Arid Environments. 1994;**27**(1):37-48

[108] Afrianto WF. Global scientific production on stakeholder engagement in community forestry: A bibliometric analysis. Jurnal Penelitian Kehutanan Wallacea. 2025;**14**(2):1-17

[109] Wild RG, Mutebi J. Conservation through community use of plant resources. People and Plants working paper. 1996;**5**:1-45

[110] Bezeng BS, Ameka G, Angui CMV, Atuah L, Azihou F, Bouchenak-Khelladi Y, et al. An African perspective to biodiversity conservation in the twenty-first century. Philosophical Transactions B. 2025;**380**(1917):20230443

[111] Kowero GS, Campbell BM, Sumaila UR. Policies and Governance Structures in Woodlands of Southern Africa. Bogor (Indonesia): Center for International Forestry Research (CIFOR); 2003

[112] Bajracharya SB, Furley PA, Newton AC. Effectiveness of community involvement in delivering conservation benefits to the Annapurna conservation area, Nepal. Environmental Conservation. 2005;**32**(3):239-247

[113] Miryeganeh M. Epigenetic mechanisms driving adaptation in tropical and subtropical plants: Insights and future directions. Plant, Cell & Environment. 2025. DOI: 10.1111/pce.15370 1

[114] Kriticos DJ, Sutherst RW, Brown JR, Adkins SW, Maywald GF. Climate change and the potential distribution of an invasive alien plant: *Acacia nilotica* ssp. indica in Australia. Journal of Applied Ecology. 2003;**40**(1):111-124

[115] Missen ML, De Kauwe MG, Hovenden MJ. Increasing photosynthetic benefit with decreasing irrigation frequency in an Australian temperate pasture exposed to elevated carbon dioxide. Journal of Experimental Botany. 2025;**40**:erae511

[116] Dhabhai K, Batra A. Hormonal regulation impact on regeneration of *Acacia nilotica* L. a nitrogen fixing tree. World Applied Sciences Journal. 2010;**11**(9):1148-1153

[117] Rahman MS, Islam SS. Rapid multiplication by nodal segment and shoot tips through in vitro micropropagation of six mulberry (*Morus alba*) genotypes. Plant Tissue Culture and Biotechnology. 2024;**34**(2):105-114

[118] Dieng M, Mbow C, Skole DL, Ba B. Sustainable land management policy to address land degradation: Linking old forest management practices in Senegal with new REDD+ requirements. Frontiers in Environmental Science. 2023;**11**:1088726

[119] Chirwa PW, Mahamane L. Overview of restoration and management practices in the degraded landscapes of the Sahelian and dryland forests and woodlands of East and Southern Africa. Southern Forests: A Journal of Forest Science. 2017;**79**(2):87-94

[120] Sarr MS, Diallo AM, King-Okumu C. A review of public versus private reforestation programs in the Senegalese Sahel: Taking stock of realities and challenges. Restoration Ecology. 2022;**30**(5):e13582

[121] Alam SA, Starr M. Deforestation and greenhouse gas emissions associated with fuelwood consumption of the brick making industry in Sudan. Science of the Total Environment. 2009;**407**(2):847-852

[122] Sulieman HM. Exploring drivers of forest degradation and fragmentation in Sudan: The case of Erawashda forest and its surrounding community. Science of the Total Environment. 2018;**621**:895-904

[123] Yasin EH, Kamil OH, Mulyana B. Multi-temporal satellite images analysis for assessing and mapping deforestation in um Hataba Forest, South Kordofan, Sudan. Journal of Sylva Indonesiana. 2022;**5**(01):81-92

[124] Coulibaly M, Bredu SA, Diwediga B, Traore SS, Yao NA. Historical changes in $CO_2$ emissions and removals from land use and land cover changes in Sudan Savannah ecological zone of Ghana. Journal of Environment and Ecology. 2020;**11**(2):1

[125] Mustafa HF, Wali BRI. Estimation of carbon dioxide assimilated in shoot biomass of mature *Acacia seyal* (Del.) Brenan and *Eucalyptus microtheca* (F.) Muell. Tropical Plant Research. 2020;7:14-19

[126] Istomo I, Farida NE. Potensi simpanan karbon di atas permukaan tanah tegakan *Acacia nilotica* L. (Willd) ex. Del. di Taman Nasional Baluran, Jawa Timur. Jurnal Pengelolaan Sumberdaya Alam dan Lingkungan (Journal of Natural Resources and Environmental Management). 2017;**7**(2):155-162

[127] Nawaz MF, Mazhar K, Gul S, Ahmad I, Yasin G, Asif M, et al. Comparing the early stage carbon sequestration rates and effects on soil physico-chemical properties after two years of planting agroforestry trees. Journal of Basic & Applied Sciences. 2017;**13**:527-533

[128] Wuepper D, Crowther T, Lauber T, Routh D, Le Clec'h S, Garrett RD, et al. Public policies and global forest conservation: Empirical evidence from national borders. Global Environmental Change. 2024;**84**:102770

[129] Kobbail AAR, Elfeel AA. Local communities involvement in forest policy implementation: Case from Sudan. Asian Journal of Agricultural Sciences. 2012;**4**(2):140-144

[130] Abdalla M, Qarmout T. An analysis of Sudan's energy sector and its renewable energy potential in a comparative African perspective. International Journal of Environmental Studies. 2023;**80**(4):1169-1187

[131] Omer AM, Braima E. Renewable energy potential in Sudan. International Energy Journal. 2001;**2**(1):15-36

[132] Adiguzel-Zengin AC, Zengin G, Kilicarislan-Ozkan C, Dandar U, Kilic E. Characterization and application of *Acacia nilotica* L. as an alternative vegetable tanning agent for leather processing. Fresenius Environmental Bulletin. 2017;**26**(12):7319-7326

[133] Prasad N, Thombare N, Sharma SC, Kumar S. Gum Arabic–A versatile natural gum: A review on production, processing, properties and applications. Industrial Crops and Products. 2022;**187**:115304

[134] Mahgoub EF, Pieper RD, Ortiz M. Use of leader lengths and diameters to estimate production and utilization of *Cercocarpus breviflorus*. Rangeland Ecology & Management/Journal of Range Management Archives. 1988;**41**(2):153-155

[135] Pandey RP, Pandey A, Galkate RV, Byun HR, Mal BC. Integrating hydro-meteorological and physiographic factors for assessment of vulnerability to drought. Water Resources Management. 2010;**24**:4199-4217

[136] Nambiar ES, Harwood CE, Kien ND. Acacia plantations in Vietnam: Research and knowledge application to secure a sustainable future. Southern Forests: a Journal of Forest Science. 2015;**77**(1):1-10

[137] Hu Y, Zhao J, Li Y, Tang P, Yang Z, Zhang J, et al. Biomass and carbon stock capacity of Robinia pseudoacacia plantations at different densities on the loess plateau. Forests. 2024;**15**(7):1242

[138] Chau MH, Quy NV, Hung BM, Xu XN, Cuong LV, Ngoan TT, et al. Soil nitrogen storage and associated regulation factors in an acacia hybrid plantation chronosequence in Southern Vietnam. Applied Ecology & Environmental Research. 2024;**22**(1):145-162. DOI: 10.15666/aeer/2201_145162

[139] Du S, Zheng B, Lei H, Guo H, Li X. Response of understory plant diversity to edge effects in plantation forests on the loess plateau. Forests. 2025;**16**(1):87

[140] Adhikari D, Singh PP, Tiwary R, Barik SK. Forest carbon stock-based bioeconomy: Mixed models improve accuracy of tree biomass estimates. Biomass and Bioenergy. 2024;**183**:107142

[141] Mouafik M, Chakhchar A, Fouad M, El Aboudi A. Remote sensing technologies for monitoring Argane forest stands: A comprehensive review. Geographies. 2024;**4**(3):441-461

[142] Guzmán ER, Rodríguez AG, Garnica JG, Munguía SM. Aplicación de sensores remotos e inteligencia artificial en la gestión y conservación de bosques frente al cambio climático en México [Application of remote sensors and artificial intelligence in the management and conservation of forests in the face of climate change in Mexico]. e-CUCBA. 2024;**1**(21):142-149

[143] El Bilali H, Rokka S, Calabrese G, Borelli T, Grazioli F, Tietiambou SRF, et al. Conservation and promotion of neglected and underutilized crop species in West Africa: Policy and governance. Sustainability (MDPI). 2024;**16**:1-21

[144] Sinthumule NI. An analysis of policy and legal framework on wetland conservation and management in South Africa. Watershed Ecology and the Environment. 2024;**6**:63-69

[145] Fernandes S, Athayde S, Harrison I, Perry D. Connectivity and policy confluences: A multi-scalar conservation approach for protecting Amazon riverine ecosystems. Perspectives in Ecology and Conservation. 2024

[146] Ferraro G, Failler P. Understanding the "implementation gap" to improve biodiversity governance: An interdisciplinary literature review. Journal of Sustainability Research. 2024;**6**(2):e240009

[147] Chimene AF, Julien HM, Sidoine MD. Effect of saline stress on the growth and physiological behavior of young planting *Acacia nilotica* in nursery and after transplantation. Plant. 2024;**12**(4):131-141. DOI: 10.11648/j.plant.20241204.16

[148] Hernandez JO, Cortes DT, Park BB. Research geographical distribution, strategies, and environmental and socioeconomic factors influencing the success of land-based restoration: A systematic review. Sustainability. 2024;**16**(15)

[149] Villarreal-Rosas J, Brown CJ, Andradi-Brown DA, Domínguez R, Jacobo P, Martínez A, et al. Integrating socioeconomic and ecological data into restoration practice. Conservation Biology. 2024;**8**(4):e14286

[150] Yue Y, Wang L, Brandt M, Zhang X, Wang K. A social-ecological framework to enhance sustainable reforestation under geological constraints. Earth's Futures. 2024;**12**(5):e2023EF004335

Section 2

# Forest Health, Monitoring and Technology

Chapter 3

# Assessing Forest Health of Sanjay Gandhi National Park, Mumbai, Maharashtra, India

*Ravindra Medhe and Chaitanya Kolekar*


## Abstract

Forest health is a crucial factor in the growth of forests. A healthy forest always grows tremendously. Forest health determines its productivity, structure, resilience, composition, process and function. There is a need to study the health of forests in Sanjay Gandhi National Park, which is encompassed by Mumbai's metro city. Sanjay Gandhi National Park is one of Maharashtra's most famous and influential national parks. This study used satellite imagery data from sentinel-2A, which has a 10-metre spatial resolution with 13 bands. Four different indices have been calculated to perform weighted overlay analysis. Out of which, three indices, NDVI, GNDVI and Modified Soil Adjust Vegetation Index (MSAVI), are used to perform weighted overlay analysis, and Moisture Stress Index (MSI), a hydrological index, is used to define the moisture stress to relate with the health of the forest. Sanjay Gandhi National Park (SGNP) is categorized into three types according to their health status: low-healthy forests, healthy forests and very healthy forests. Low-healthy forests acquired 39.52%, healthy forests acquired 36.51% and very healthy forests acquired 3.36% of the total area of Sanjay Gandhi National Park.


**Keywords:** forest health, weighted overlay, vegetation index, Sanjay Gandhi National Park, remote sensing, remote sensing, GIS

## 1. Introduction

Forest health is a condition that helps to regulate the structure, composition, processes, function, productivity and resilience of forest ecosystems over time and space [1]. Forests get disturbed worldwide due to urbanization, industrialization, illegal mining, overharvesting, habitat degradation and other unsustainable management practices [2]. Health is a very crucial factor in any forest. It determines the effect on the environment. A healthy forest always makes its surroundings healthy, and a healthy environment keeps the forest healthy. Forest health assessment is essential to determine appropriate measures for sustainable forest ecosystems [3]. Healthy forests are more stable and vigorous in response to different stresses, disturbances and resource restrictions [4]. As technology advances day by day, remote sensing (RS), geographical information systems (GIS) and global positioning systems (GPS) are becoming practical tools for monitoring and determining forest health [5]. Although the forests experienced drought, fires, disease outbreaks and pollution, they have evolved worldwide. They have also fought artificial disturbances and stresses that affect the condition of forests, either directly through cutting and clearing or indirectly through invasion of exotic species, pollution and climate change [6]. Forest area in the whole world, as a proportion of total land area, decreased from 32.5 to 30.8% in the last three decades between 1990 and 2020. This depicts a net loss of 178 million hectares of forest, which is about equal to the size of Libya. Forest areas were reduced due to the expansion of agricultural land and increased due to the natural expansion of forests. Sixty-seven million hectares of forest burned annually between 2003 and 2012, and 35 million hectares of forest area were damaged due to the outbreaks of forest insect pests [7].

The National Forest Policy of India 1988 deliberates on achieving a goal of 33% of the country's geographical area covered by a forest. The forest cover includes lands of more than 1 hectare in area and has a tree canopy density of more than 10%. The total forest cover in India is 21.71%, out of which only 3.04% is a very dense forest that has a tree canopy density of more than 70%, 9.33% is a moderately dense forest which has 40 to 70% of tree canopy density, and 9.34% is an open forest that has tree canopy density between 10 and 40%. Maharashtra is India's 5th largest state in forest cover, with 16.51% of land covered by forest [8]. This study is based on four indices, NDVI, GNDVI, MSAVI and MSI, which are useful in vegetation analysis. Each index has its specific values that help analyze forest health. An integrative study of these indices is performed through a weighted overlay by giving the appropriate weights to each index. Weighted overlay has been done on only three indices except MSI, a hydrological index. Each index carries equal weight because all these three indices are equally crucial to detecting forest health.

Sentinel-2A satellite data is utilized to assess the health of the SGNP forest; sentinel-2A's multispectral imager can capture data across 13 bands, providing vital information on the vegetation's state. SGNP is one of Maharashtra's most famous and influential national parks. It is situated within Mumbai and is the only national park encompassing the world's metropolitan city. There is a need to study the health of various forests in Maharashtra, a very underrated research subject. However, it gives a proper way to protect these forests with the help of GIS techniques. The main objective of this study is to determine the forest health of SGNP, which can be consummate by differentiating between forest types based on a canopy cover.

## 2. Review of literature

Spatiotemporal forest health assessment for Sariska National Park (SNP) India, ecosystem management under regional climatic inconsistencies. This study was conducted in SNP to determine the forest health and changes from 1972 to 2018. This study used satellite datasets of Landsat and Sentinel-2, combining data of some climatic parameters. Indices, such as NDVI, MSI, DVI and MSAVI-2, were calculated to find the relation between rainfall and these indices. This study shows the changes in forest cover in the past 46 years by representing the dense forest and open forest cover in the percentage. This study also represented the change detection of agriculture, settlement and temperature during the study period. Overall, this study depicts changes in the health of the forest concerning climatic parameters [9].

Aigbokhan, John worked on forest health analysis using remote sensing and GIS techniques. A Case Study of Omo Forest Reserve, Ogun State, Nigeria. This study was conducted to assess the health status of the Omo Forest reserve. They used a satellite image of Landsat 8 (OLI-TIR) to perform vegetation indices. These indices were used to process the weighted sum overlay analysis further to determine the health status of the forest. This study shows the map with four reclassified categories, that is, non-forest, unhealthy, moderately healthy and very healthy. The result of the study indicates an area of these reclassified categories in percentage: non-forest covered about 6%, unhealthy forest covered 25%, moderately healthy forest spread in about 44% and very healthy forest covered about 25% of the study area [10].

Roshani et al. researched assessing forest health using remote sensing-based indicators and fuzzy analytic hierarchy process in Valmiki Tiger Reserve, India. This study is based on an assessment of the forest health of Valmiki Tiger Reserve. They derived some datasets from remote sensing data, such as vegetation indices, forest fragmentation, rainfall and soil types. Vegetation indices like advanced vegetation index, normalized difference vegetation index and normalized difference moisture index were calculated and multiple buffer zones for various features were also created. The final interpretation of these data was done using a fuzzy analytic hierarchy process. These layers were used to create a forest health map with the help of the weighted overlay method. The result indicates that the forest cover area is divided into various categories. Non-forest areas covered about 19%, unhealthy forests covered 13% and moderately healthy forests, which were found to be the largest forest area, covered about 37% of the total study area [11].

Larasati et al. studied utilizing the normalized difference vegetation index (NDVI) transformation for healthy forest assessment. This study was conducted in Nanggala III National Park, a nature conservation area, and it aims to assess the level of health of the forest in Nanggala III National Park. This study uses the transformation of vegetation index and field observation, while the purposive sampling method was used. Indicators for assessing forest health level using analysis of vegetation index (NDVI) using Landsat 8, recorded in 2015. The forest area in Nanggala III Natural Park is categorized as a healthy forest. The vegetation densities in the medium with an area of 97.74 ha (10.09%), sparse vegetation densities in the area of 26.28 ha (2.71%) and non-vegetation cover (cloud and build area) amounted to 45.72 ha (4.72%) [12].

## 3. Datasets and tools

To analyze the forest health of SGNP, satellite data of sentinel-2A were downloaded from the Google Earth Engine website (https://code.earthengine.google.com). Sentinel-2A is a satellite that provides high-resolution images of the Earth's surface, including vegetation and is helpful for vegetation analysis. Sentinel-2A's multispectral imager can capture data across 13 different bands, including three in the "red edge", which provide vital information on the state of vegetation. The greenness component can show variations in photosynthetically active vegetation, and the red-edge band is linked to the chlorophyll content of leaves. The shortwave infrared (SWIR) bands can help identify variations in vegetation stress and moisture content. ArcGIS Pro performs the different indices using the spatial analyst tool, the raster calculator. A weighted overlay of these indices has also been done using this software.

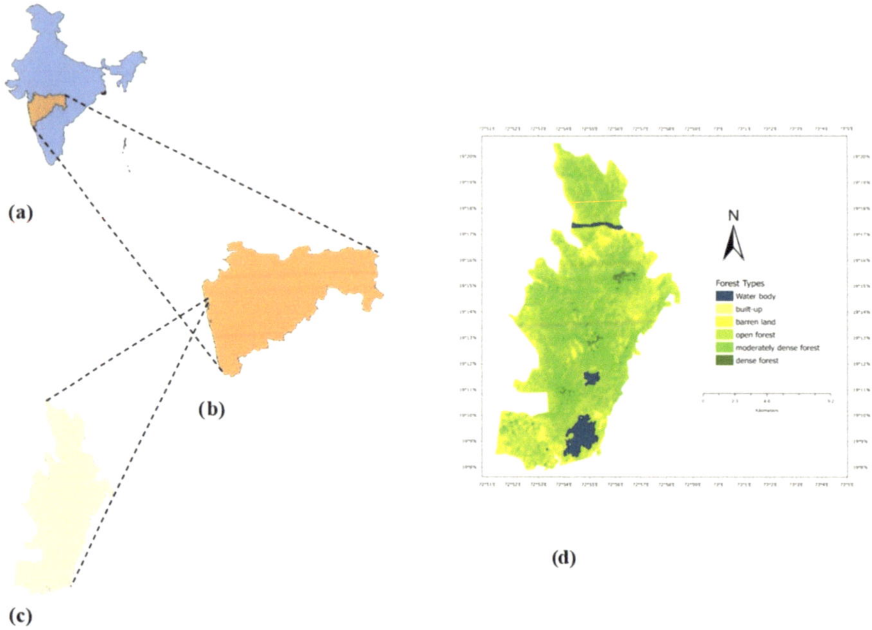


**Figure 1.**
*Study area: (a) India, (b) Maharashtra, (c) SGNP and (d) Forest types – SGNP.*

## 4. Study area

The SGNP is between 72° 53′ and 72° 58′ East longitude and 19° 88′ to 19° 21' North latitude and lies within the city limits of Mumbai. Mumbai is the world's second-largest metropolitan city and the only city that enwinds a National Park. The total area of the National Park is 103 km$^2$ [13]. The SGNP has categorized various forest types, containing 82.39 km$^2$ of reserved forest, 0.21 km$^2$ of protected forest and 20.76 km$^2$ of un-classed forest. The SGNP is also divided into different zones according to the importance of the forest, that is, core zone (CZ) of 76.92 km$^2$, Eco-Tourist Zone (Eco-TZ) of 0.45 km$^2$, Buffer Zone (BZ) or multiple use zone of 5.89 km$^2$ (**Figure 1**) [14].

## 5. Methodology

See **Figure 2**.

### 5.1 Vegetative indices

Vegetative indices are used to depict the greenness and forest health of SGNP. These are four indices: NDVI (Normalized Difference Vegetation Index) is a ratio of the red and NIR band that shows the greenness of a region, and its value range lies between −1 and + 1. MSI (Moisture Stress Index) is a hydrological index that specifies water availability and is helpful to show forest health. The MSAVI (Modified Soil Adjust Vegetation Index) is the vegetative index, which minimizes the bare soil effect on vegetation and has a range value between −1 and + 1 [9]. GNDVI shows the chlorophyll variation of plants through the greenness of plants:

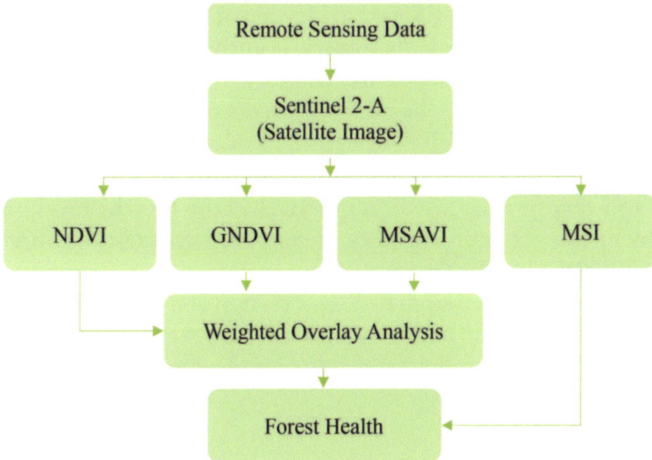


**Figure 2.**
*Methodological Framework.*

$$\text{NDVI} = (NIR + RED)/(NIR - RED) \tag{1}$$

$$\text{GNDVI} = (NIR - \text{GREEN})/(NIR + \text{GREEN}) \tag{2}$$

$$\text{MSAVI} = \left(2^{*}\ NIR + 1 - \text{sqrt}\left(\left(2^{*}\ NIR + 1\right)^{2} - 8^{*}\left(NIR - RED\right)\right)\right)/2 \tag{3}$$

$$MSI = \text{SWIR}/NIR \tag{4}$$

### 5.2 Integrative analysis of vegetative indices

This study is based on four different indices useful in vegetation analysis. Each index has its specific values that help analyze forest health. An integrative study of these indices is performed through a weighted overlay by giving the appropriate weights to each index. Weighted overlay has been done on only three indices except MSI, a hydrological index. Each index carries equal weight because all these three indices are equally crucial to detecting forest health.

## 6. Results

### 6.1 Forest health status

Forest health is a broad term, as a forest contains not only trees, bushes and grasses but also the animals and waterbodies within it. So, a healthy forest is just a reflection of healthy surroundings. MSI (Moisture Stress Index) gives us a comprehensive view of moisture availability, which is an essential factor in the growth of the forests. Less moisture stress depicts the healthy growth of vegetation, and more moisture stress creates difficulty in vegetation growth. All these three vegetation indices (NDVI, GNDVI and MSAVI) are vital in determining forest health. The weighted overlay of these indices provides an appropriate status for forest health. The range from 0.5 to 0.6 represents an open forest (39.52%, **Figure 3**) with less tress density. Range from 0.6 to 0.7 represented a moderately dense forest (36.51%, **Figure 4**) with a moderate density of trees. A range of more than 0.7 represents a dense forest (3.36%, **Figure 5**), which also depicts the forest's high density and healthy growth (**Figure 6**). Open forest, moderately dense forest and dense forest is later described as low-healthy forest, healthy forest and very healthy forest, respectively. Low-healthy forests acquired 39.52%, healthy forests acquired 36.51% and very healthy forests acquired 3.36% of the total area of Sanjay Gandhi National Park.

Forest health is directly related to its density. High density shows healthy forest growth and less density shows minimum growth. So, we can conclude that dense forest is a very healthy forest, followed by moderately dense forest and open forest, which are less healthy.

### 6.2 Vegetative indices

This study was conducted with the help of four indices. Of these four, three indices, that is, NDVI, GNDVI and MSAVI, are vegetative and MSI is a hydrological index.

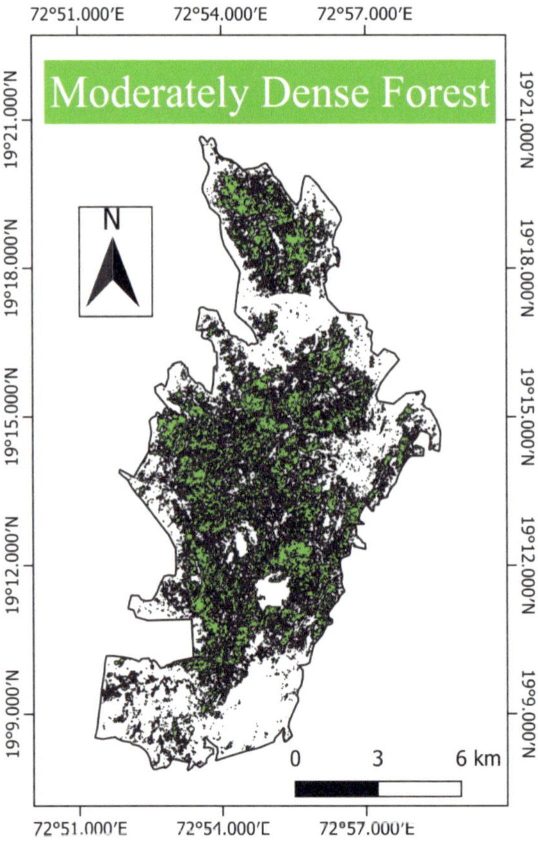


**Figure 3.**
*Open Forest.*

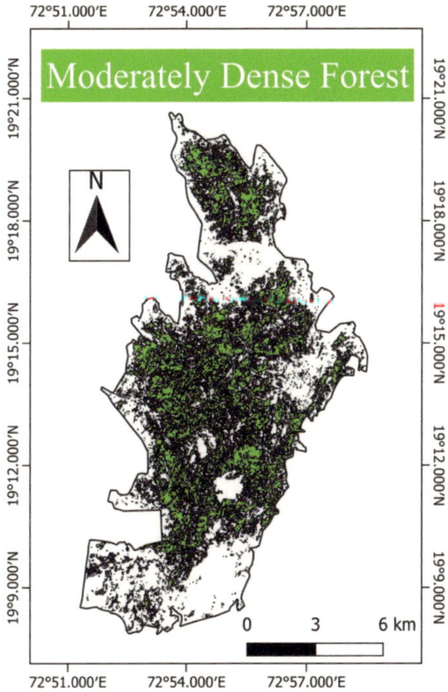


**Figure 4.**
*Moderately Dense Forest.*

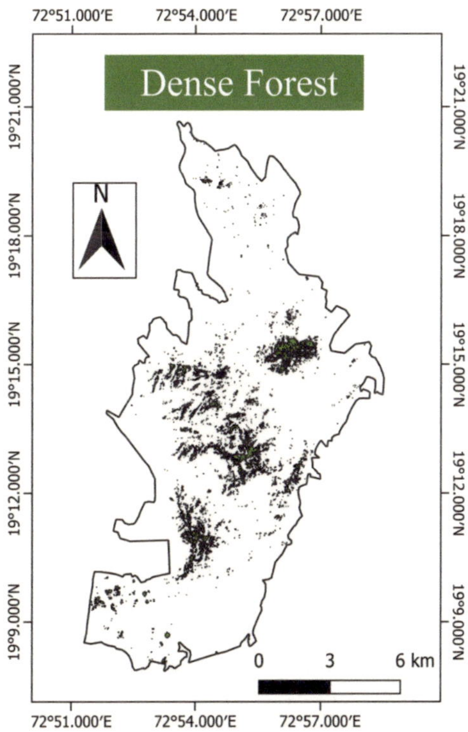


**Figure 5.**
*Dense Forest.*

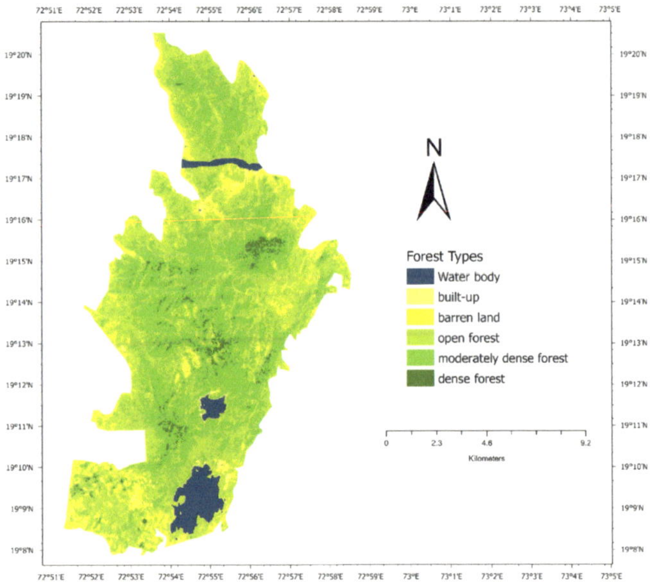


**Figure 6.**
*Weighted Overlay.*

### *6.2.1 Normalized difference vegetative index (NDVI)*

NDVI is calculated by measuring the difference between near-infrared (NIR) and red light. Vegetation strongly reflects NIR but absorbs or has low reflectance of red light. NDVI ranges from −1.0 to 1.0 (**Figure 7**).

Generally, the values less than 0 indicate water bodies. Very low values, such as 0.1 or less, indicate barren lands. Range from 0.2 to 0.5 indicates sparse vegetation. Dense vegetation ranges from 0.6 and above.

### *6.2.2 Green normalized difference vegetation index (GNDVI)*

GNDVI (Green Normalized Difference Vegetation) is a vegetative index indicating plants' greenness or photosynthetic activity. It is more sensitive to chlorophyll variation in the crop than NDVI and has a higher saturation point (**Figure 8**).

Generally, the range between −1 and 0 represents the water bodies, bare soil and rocks. Values greater than 0 indicate the vegetation—the more intense the green, the denser the vegetation cover.

### *6.2.3 Modified soil adjusted vegetation index (MSAVI)*

MSAVI is a vegetation index which minimizes the bare soil effect on the reflectance of vegetation. Usually, it is used in areas covered mainly by bare soil or rock, but it can also be used in normal vegetation areas (**Figure 9**).

The range between −1.0 to 0.2 indicates the water and bare soil. Seed germination and leaf development stages are indicated between 0.2 to 0.4 and 0.4 to 0.6, respectively. A range greater than 0.6 indicates dense vegetation enough to cover the soil. So, we conclude here that the forest in SGNP is mature and moderately dense to dense because most of the areas show values greater than 0.6.

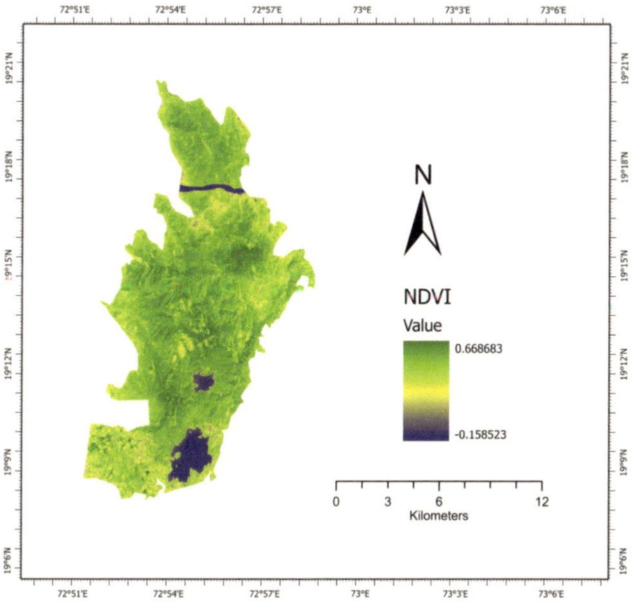


**Figure 7.**
*Normalized Difference Vegetation Index.*

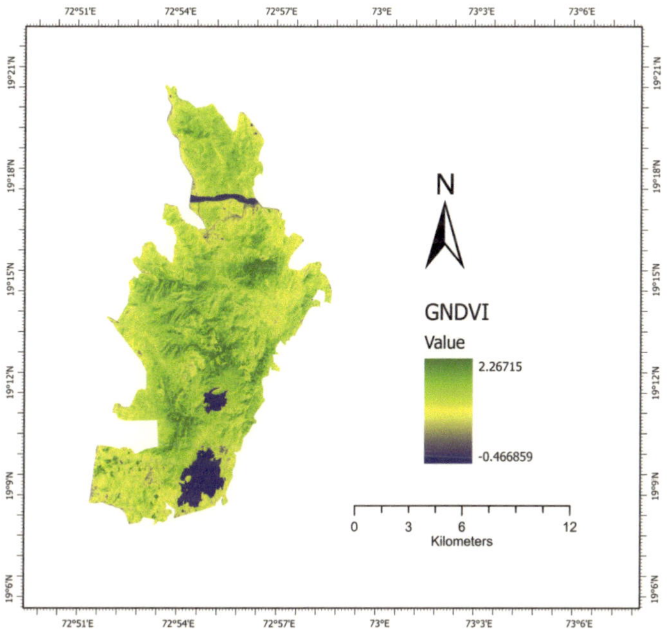


**Figure 8.**
*Green Normalized Difference Vegetation Index.*

### *6.2.4 Moisture stress index (MSI)*

MSI is a combination of vegetative and hydrological indices which depict the water content of leaves in vegetation. It shows the moisture stress in a given area (**Figure 10**).

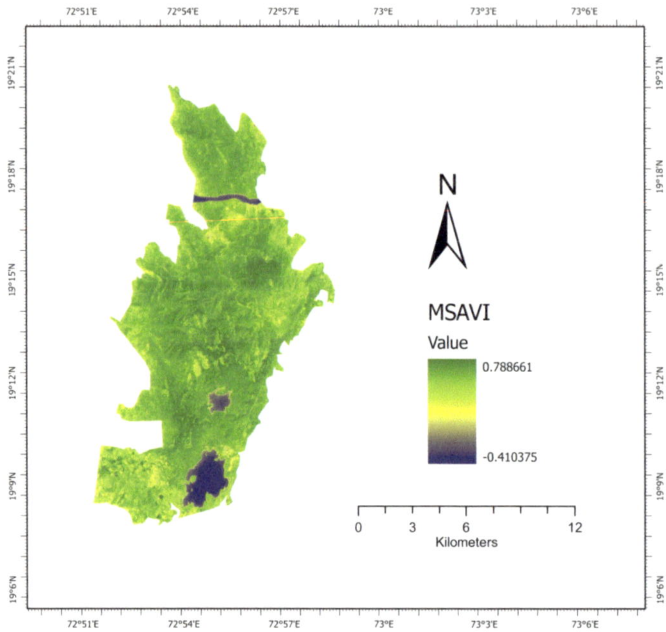


**Figure 9.**
*Modified Soil Adjusted Vegetation Index.*

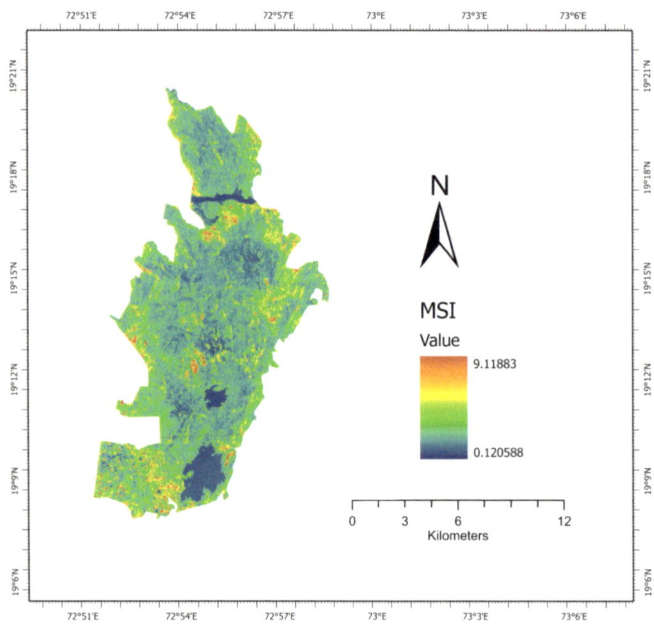


**Figure 10.**
*Moisture Stress Index.*

Unlike other vegetation indices, this index is a reverse index because higher values show high moisture stress and low water content, adversely affecting vegetation growth.

## 7. Conclusion

Forest ecosystems play a vital role in human life. Nevertheless, speedy industrialization, urbanization and vehicle increases adversely affect the forest ecosystem. So, a forest health assessment is necessary to achieve the goal of ecologically sustainable development. Forest areas near metropolitan cities, like SGNP, are affected more and earlier and are surrounded by three cities: Mumbai, Thane and Vasai-Virar. This study delineated these effects on the forest of SGNP through various indices and their weighted overlay analysis. Weighted overlay of NDVI, GNDVI and MSAVI has been categorized into three forest types according to their canopy cover: open forest, moderately dense forest and dense forest, which is later described as low-healthy forest, healthy forest and very healthy forest, respectively. Low-healthy forests acquired 39.52%, healthy forests acquired 36.51% and very healthy forests acquired 3.36% of the total area of Sanjay Gandhi National Park. The non-forest area includes built-up, barren land and waterbodies acquired 20.32% of the total area. Moisture stress index (MSI) shows low moisture stress in healthy and healthy forests, medium moisture stress in forest areas of low-healthy forests and high moisture stress in barren land and built-up areas. So, this study also concluded that moisture and water availability can directly affect the health of vegetation.

## Abbreviations

| | |
|---|---|
| DVI | Differentiate Vegetation Index |
| FAO | Food and Agriculture Organization |
| FSI | Forest Survey of India |
| GIS | Global Information System |
| GNDVI | Green Normalized Difference Vegetation Index |
| GPS | Global Positioning System |
| MSAVI | Modified Soil Adjust Vegetation Index |
| MSI | Moisture Stress Index |
| NDVI | Normalized Difference Vegetation Index |
| OLI-TIR | Operational Land Imager and Thermal Infrared |
| SNP | Sariska National Park |
| SGNP | Sanjay Gandhi National Park |
| SWIR | Shortwave Infrared |
| UNEP | United Nations Environment Programme |

## Author details

Ravindra Medhe* and Chaitanya Kolekar
Department of Geography, Savitribai Phule Pune University, Pune, Maharashtra, India

*Address all correspondence to: ravindramedhe1@gmail.com

## References

[1] NASEM. National academies of sciences, engineering, and medicine. In: Forest Health and Biotechnology: Possibilities and Considerations. Washington, D.C.: National Academies Press; 2019

[2] FAO. Global forest research assessment. In: Main Report. Rome, Italy: FAO Forestry Paper; 2010

[3] Meng Y, Cao B, Dong C, Dong X. Mount Taishan forest ecosystem health assessment based on forest inventory data. Forests (MDPI). 2019;**10**(8):1-14, 657. DOI: 10.3390/f10080657

[4] Bera B, Saha S, Bhattacharjee S. Estimation of Forest Canopy Cover and Forest Fragmentation Mapping Using Landsat Satellite Data of Silabati River Basin (India). New York, USA: Springer Science and Business Media LLC; 2020

[5] Rahaman MH, Sajjad H, Roshani MM, Bhuyan N. Rehman S. Delineating Groundwater Potential Zones Using Geospatial Techniques and Fuzzy Analytical Hierarchy Process (FAHP) Ensemble in the Data-Scarce Region: Evidence from Manipur's Lower Thoubal River Watershed. New York, USA: Springer Science and Business Media LLC; 2022

[6] Trumbore S, Brando P, Hartmann H. Forest health and global change. Science. 2015;**349**(6250):814-818

[7] FAO and UNEP. The state of the World's forests 2020. In: Forests, Biodiversity and People. Rome, Italy: Food and Agriculture Organization of the United Nations (FAO) and the United Nations Environment Programme (UNEP); 2020

[8] FSI. India State of Forest Report. Dehradun, Uttarakhand, India: The Forest Survey of India (FSI), which operates under the Ministry of Environment, Forest and Climate Change, Government of India; 2021

[9] Sharma L, Raj A. Spatiotemporal Forest Health Assessment for Ecosystem Management of Sariska National Park (India) under Regional Climatic Inconstancies. Dordrecht, Netherlands: Springer Nature B.V.; 2024

[10] Aigbokhan JO. Forest Health Analysis, Using Remote Sensing and GIS Techniques: A Case Study of Omo Forest Reserve, Ogun State, Nigeria: Science and Education Development Institute; 2018

[11] Roshani, Sajjad H, Rahaman MH, Rehman S, Masroor M, Ahmed R. Assessing forest health using remote sensing-based indicators and fuzzy analytic hierarchy process in Valmiki Tiger Reserve, India. International Journal of Environmental Science and Technology. 2023;**20**:8579-8598

[12] Larasati D, Setiawan H, Muzayanah. The Utilization of Transformation of Normalized. Paris, France: Atlantis Press; 2018

[13] Bhale P, Bhatti I, Mayes R. Sanjay Gandhi National Park (SGNP) A Concise Report. Self-published via ResearchGate; 2017

[14] Pande P. National Parks and Sanctuaries in Maharashtra Reference Guide – Volume I and II. Mumbai: Bombay Natural History Society; 2005

Chapter 4

# Utilization of Remote Sensing Technology and IoT for Forest Condition Monitoring and Environmental Analysis

*Asep Denih*


## Abstract

Forest monitoring is essential for environmental conservation and resource management, addressing deforestation, climate change, illegal activities, and natural disasters. This book chapter explores the application of remote sensing technology and the Internet of Things (IoT) for accurate, efficient, and sustainable forest condition monitoring. Remote sensing technologies, such as Landsat, MODIS, and Sentinel, track land cover changes, vegetation health, and illegal activities, while vegetation indices like NDVI and EVI assess ecosystem conditions. IoT technology, with real-time sensors measuring temperature, humidity, soil moisture, and air quality, provides continuous data for early risk detection. This integrated approach combines large-scale data collection with real-time analysis, empowering stakeholders to address forest challenges effectively. The integration of IoT with Geographic Information Systems (GIS) enhances spatial analysis, supporting informed decision-making for forest conservation. By adopting these innovations, forest ecosystems can be better preserved, ensuring sustainability and resilience against environmental threats.


**Keywords:** forest monitoring, remote sensing, internet of things (IoT), geographic information systems (GIS), vegetation indices

## 1. Introduction

Monitoring the state of forests and their ecosystems means protecting the sustainability of the area as one of the components of the ecosystem, which is very important. Forests have a main role in absorbing carbon, protecting air balancers, providing habitats for various flora and fauna, and supporting human life with different natural energy sources. However, threats to forests, such as deforestation, climate change, illegal activities, and natural disasters, continue to increase, so efficient and effective monitoring procedures are needed [1].

Modern technology has introduced new solutions to the challenges of forest monitoring, namely through remote sensing technology and the internet of things (IoT). Remote sensing technology can collect regional data remotely without requiring a physical presence in the field. By utilizing sensors installed on satellites or aircraft, this technology is able to generate various ecological parameters, such as green land area, forest degradation rate, and land cover change [2]. This technology can also monitor forest conditions periodically to create early changes. Using satellite imagery from a variety of sources, such as Landsat, MODIS, and Sentinel, he shared accurate information and covered a wide zone.

On the other hand, IoT technology [1] opens up opportunities for real-time forest monitoring by using a network of sensor features connected to IoT sensors that can measure various area parameters such as temperature, humidity, and air quality. Information collected from IoT sensors can be accessed directly through the internet network to better respond to changes in forest conditions [3]. Recent advancements in IoT technology include the use of low-power wide-area networks (LPWAN) and satellite communication, enabling sensors to operate efficiently in remote forest regions with limited connectivity. These innovations significantly enhance the reliability and scalability of IoT-based forest monitoring systems.

IoT can also use bonus sensors, such as sound and cameras, to find human activities or wild animals in the forest. The integration of IoT technology with geographic information systems (GIS) strengthens the ability to visualize location-based information [2]. Furthermore, the integration of artificial intelligence (AI) with IoT and GIS has shown great potential in predictive forest risk management, such as identifying patterns that indicate possible deforestation or wildfires. These capabilities provide stakeholders with valuable tools for proactive forest conservation. This approach provides a deeper insight into the distribution of regional conditions within the forest which helps in analyzing more comprehensively.

Global efforts, including the use of open data platforms and international collaboration, have further amplified the impact of these technologies. These initiatives aim to make forest monitoring systems more accessible and effective for conservation efforts worldwide.

This book chapter explains in depth the use of remote sensing technology and IoT for forest monitoring. In the discussion, the types of sensors used, how this technology works, and its effectiveness in supporting conservation efforts are explained. Not only that, the review also includes analytical procedures used to evaluate forest health, such as the Normalized Vegetation Index (NDVI) and Enhanced Vegetation Index (EVI), which can take into account the state of vegetation periodically [3]. Through this technology, monitoring the state of forests can be tried more accurately, effectively, and efficiently to support appropriate decision-making to conserve and manage forests in a sustainable manner [4].

## 2. Remote sensing technology for forest monitoring

Remote sensing technology is an important tool in monitoring forest conditions. This technology can collect data about changes and forest conditions without going directly to the site. Remote sensing involves using sensors that are usually placed on satellites or airplancs to monitor various environmental parameters, such as the area of green land, the level of damage caused by fires, changes in land cover, and the level of forest health [5].

The information generated from remote sensing is very useful for various purposes, such as detecting illegal logging activities, monitoring plant health, and considering the consequences of climate change on forest ecosystems. Thus, the data obtained can be used to make more appropriate decisions in forest conservation and natural energy resource management [6].

The advantage of this technology is its ability to provide accurate data, cover a large area quickly, and collect data in hard-to-reach locations. In addition, remote sensing can also carry out periodic monitoring so that changes that occur in the forest can be detected more quickly and efficiently. This is certainly very important in efforts to protect forests and mitigate environmental disasters [3].

### 2.1 Sensor types and satellite imagery

Remote sensing utilizes sensors installed on satellites to collect information about the environment, including forests. These sensors are divided into two main categories: optical and radar sensors. Both have different characteristics and functions in collecting environmental data [4].

1. Sensor optics

   Optical sensors use sunlight reflected by the Earth's surface to collect data. These sensors work on a wide range of wavelengths, from visible light to infrared [3]. Some satellites that use optical sensor technology include the following:

a. Landsat

   The Landsat satellite series has operated since 1972 and is often used in remote sensing. The satellite is equipped with a multispectral sensor capable of recording information from a portion of the wavelength range. Information from Landsat is useful for observing land change, vegetation monitoring, forest health, and finding land cover changes over time [6].

b. MODIS (Moderate Resolution Imaging Spectroradiometer)

   These sensors are attached to NASA's Terra and Aqua satellites. MODIS can record data with a wide coverage daily for more frequent monitoring. MODIS data are used for large-scale analyses such as monitoring forest fires, climate change, and vegetation distribution [5].

c. Sentinel

   As part of the Copernicus program managed by the European Space Agency (ESA), the Sentinel satellite has several variants with different types of sensors. For example, Sentinel-2 is equipped with a multispectral optical sensor capable of providing high resolution for environmental monitoring, such as vegetation changes, agriculture, and forest health [7].

2. Sensor radar

   Unlike optical sensors, radar sensors do not rely on sunlight to work. These sensors utilize microwaves emitted from satellites and reflected to collect data. This technology is particularly effective for observation in adverse weather conditions or at night [8]. Some satellites that use radar sensors include the following:

a. Sentinel-1

As part of the Copernicus program, Sentinel-1 uses Synthetic Aperture Radar (SAR) technology to produce high-quality imagery regardless of weather or lighting conditions. These data are very useful for monitoring deforestation, soil movement, and soil moisture conditions [6].

b. RADARSAT

Developed by Canada, RADARSAT uses radar technology capable of detecting changes on the Earth's surface. RADARSAT imagery is very useful for forest mapping, analysis of land change patterns, and identification of climate change and natural disasters [5].

### 2.2 Vegetation analysis with vegetation index (NDVI/EVI)

Vegetation analysis using vegetation indices such as the Normalized Difference Vegetation Index (NDVI) [9] and Enhanced Vegetation Index (EVI) [10] has become a commonly used method in remote sensing. This method plays an important role in monitoring forest health conditions, land cover changes, and forest degradation rates. These two indices can help researchers evaluate the state of vegetation extensively and periodically and provide an accurate picture of forest ecosystem dynamics [9].

Normalized Comparative Vegetation Index (NDVI) is an index used to measure vegetation health using light reflection from the plant surface. The NDVI bottom concept is a comparison between the reflection of near-infrared (NIR) rays and red visible light (Red) captured by satellite sensors. Healthy plants have a lot of chlorophyll, reflect more NIR rays, and absorb red light, creating a large NDVI value [6]. The NDVI formula is

$$NDVI = \frac{(NIR - Red)}{(NIR + Red)} \quad (1)$$

NDVI values range from −1 to 1. A value close to 1 indicates lush and healthy vegetation. In contrast, a value close to 0 or negative indicates an area with little or no vegetation, such as open land, water bodies, or areas that have experienced deforestation. Uses of NDVI include the following:

1. Forest health monitoring: NDVI is often used to observe forest health patterns, including identifying areas experiencing stress due to drought or pollution [11].

2. Land cover analysis: With NDVI, researchers can compare vegetation changes over time by identifying deforestation or other changes in land cover [6].

3. Forest degradation detection: Declining NDVI values indicate a decline in vegetation health resulting from human activities, climate change, or natural disasters such as forest fires [8].

While NDVI is very useful, the Enhanced Vegetation Index (EVI) was developed to address some limitations, especially its sensitivity to atmospheric and saturation influences in dense vegetation areas. EVI uses more advanced atmospheric correction and adds a blue light spectrum to reduce the effects of fog or dust, making it more accurate in areas with dense vegetation [3].

$$EVI = G \times \frac{(NIR - Red)}{(NIR + C_1 \times Red - C_2 \times Blue + L)} \tag{2}$$

Where

1. G is the gain factor (usually 2.5),

2. C1 and C2 are atmospheric correction coefficients,

3. L is the stabilization factor that reduces the noise from the ground background effect.

Uses of EVI:

1. Dense vegetation monitoring: EVI is particularly effective in areas with very dense vegetation, such as tropical rainforests. This index distinguishes the variation of vegetation health conditions in areas with high NIR reflection [12].

2. Seasonal change analysis: EVI plays an important role in evaluating seasonal climate change in forests, especially in areas experiencing dry and wet season cycles [13].

3. Environmental degradation studies: EVI helps researchers identify regions that show signs of degradation, both caused by human activities and natural factors [12].

### 2.3 The use of remote sensing in the analysis of lost Greenland

The use of satellite imagery, such as Landsat 8, has proven to be an effective method of analyzing land cover changes, including the loss of green areas such as Greenland. By utilizing the NDVI (Normalized Difference Vegetation Index) algorithm, researchers can accurately map vegetation cover based on the reflection of the light spectrum captured by the satellite. This analysis can identify areas with declining vegetation, which is an important indicator of the impact of urbanization, climate change, or other human activities.

In a case study in Bogor City, Python programming offers flexibility and efficiency in processing satellite image data. Libraries such as raster, NumPy, and Matplotlib speed up data processing, produce informative visual maps, and perform statistical analysis to detect vegetation trends over time. This approach not only highlights areas experiencing vegetation loss but also provides a robust framework for understanding the impacts of urbanization and supporting sustainable urban planning [14].

Remote sensing technology combined with advanced data analysis plays a crucial role in monitoring environmental changes. These technologies not only improve our ability to detect and measure the loss of green areas but also provide valuable insights into designing mitigation and adaptation strategies to face evolving environmental challenges.

The analysis shown in **Figure 1** shows the results of combining various Landsat 8 imagery channels to analyze vegetation in the observed region. By utilizing specific bands (such as Band 4, Band 5, and Band 6), the spectral properties of vegetation, clouds, and cloud shadows become distinguishable, a detailed examination of vegetation cover.

The combination of these bands, often through the NDVI or other vegetation indices, enhances the ability to identify green vegetation. In the final image, areas with healthy vegetation are shown in bright green, while areas without vegetation look darker. The presence of clouds and their shadows is an important factor in ensuring accurate vegetation analysis, as these elements can interfere with spectral reflection data. By identifying and removing clouds and their shadows precisely, researchers can focus their analysis on actual vegetation signals without experiencing distortion.

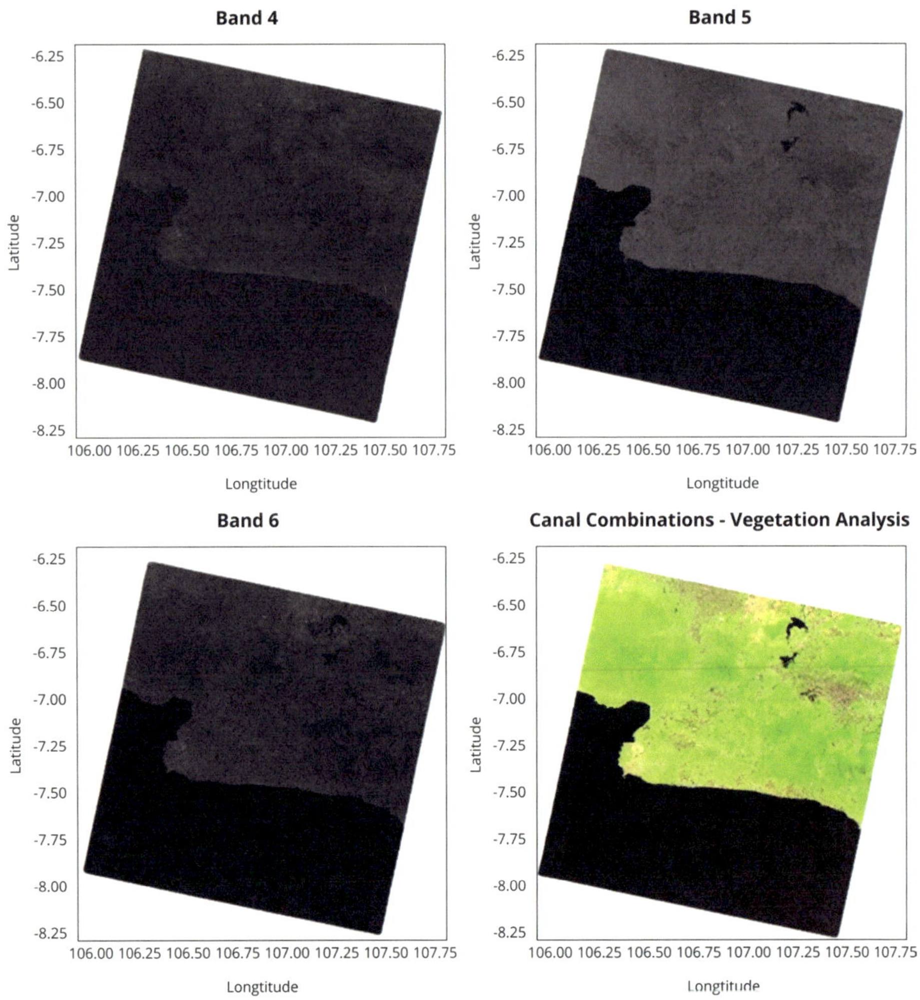


**Figure 1.**
*Results of canal combinations: vegetation analysis.*

This approach improves the accuracy of vegetation monitoring and supports a wider range of applications, such as land cover classification, environmental monitoring, and change detection over time. The results of this analysis greatly contribute to sustainable land management and conservation planning in urban and rural areas.

The analysis in **Figure 2** compares vegetation analysis images before and after the radiometric calibration process. It demonstrates that there is no noticeable change in the color brightness between the two images. This outcome is expected because the radiometric calibration process on bands 4, 5, and 6 primarily adjusts the pixel value range from 0 to 65,535 to 0–1 without altering the visual representation of the data [15].

Radiometric calibration is an important first step in remote sensing data processing to ensure the consistency and accuracy of satellite images. By standardizing pixel values, this process performs better comparisons between datasets, regardless of differences in sensor characteristics, atmospheric conditions, or lighting variations during data capture. Although the visual appearance of the image does not change, radiometric calibration ensures that the numerical data generated from the image is reliable for quantitative analysis. Fortunately, the radiometric technology of calibration in Landsat satellite imagery is currently more advanced, so this process can be handled automatically by satellite systems.

This step is very important in vegetation analysis, as it improves the accuracy of indices such as NDVI so that the resulting values truly reflect the health conditions and density of vegetation. The calibrated data provides a solid foundation for advanced analysis, monitoring of environmental changes, evaluation of vegetation dynamics, and supporting decision-making in land management and conservation efforts.

**Figure 3** shows the overlay process, in which the administrative map of Bogor City is superimposed on the Landsat 8 imagery. This step aligns administrative boundaries with the corresponding geographic coordinates of the Landsat data, thus ensuring accuracy and spatial fit.

The overlay technique is an important step in geospatial analysis because it combines different layers of data to produce more comprehensive insights. In this case, the administrative map helps to provide context to the vegetation analysis by

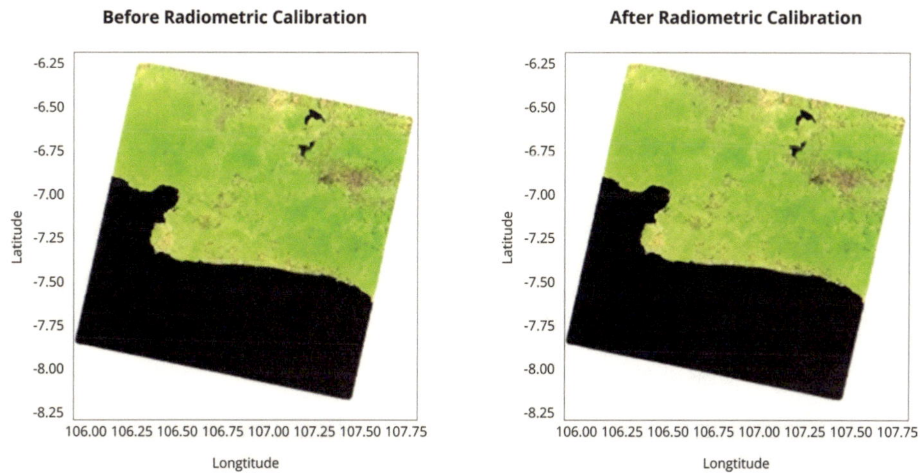


**Figure 2.**
*Comparison of vegetation analysis images before and after radiometric calibration.*

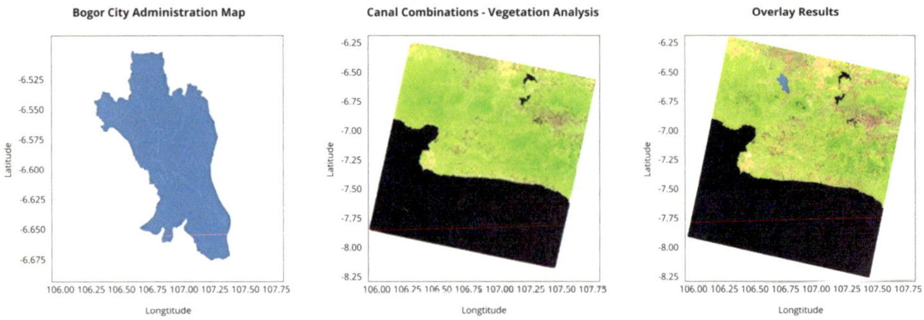


**Figure 3.**
*Cloud removal process and cloud shadows.*

clearly depicting the study area within the administrative boundaries of Bogor City. This spatial alignment can help researchers evaluate vegetation conditions in a given region, monitor land use patterns, and analyze environmental changes within an administrative framework.

The combination of satellite imagery with administrative maps is very beneficial for urban planning, environmental management, and policymaking. By integrating these data sources, stakeholders can make better decisions regarding land use, conservation, and sustainable development strategies in urban areas. This overlay approach also improves the visualization of the analysis's results, making the findings easier for various parties to understand and apply.

**Figure 4** depicts the process of overlaying and cropping the image used to focus the analysis on the Bogor City area in the Landsat 8 imagery. The original satellite image covers a significantly larger area, with Bogor City occupying only 0.32% of the entire image. To optimize computational efficiency and focus the analysis, the image is cropped to isolate the region of interest.

Before cropping, the image matrix size is 7787 x 7669, encompassing a vast amount of unnecessary data outside the city's boundaries. After the cutting process, the matrix size is reduced to 624 x 425, significantly reducing the volume of data that

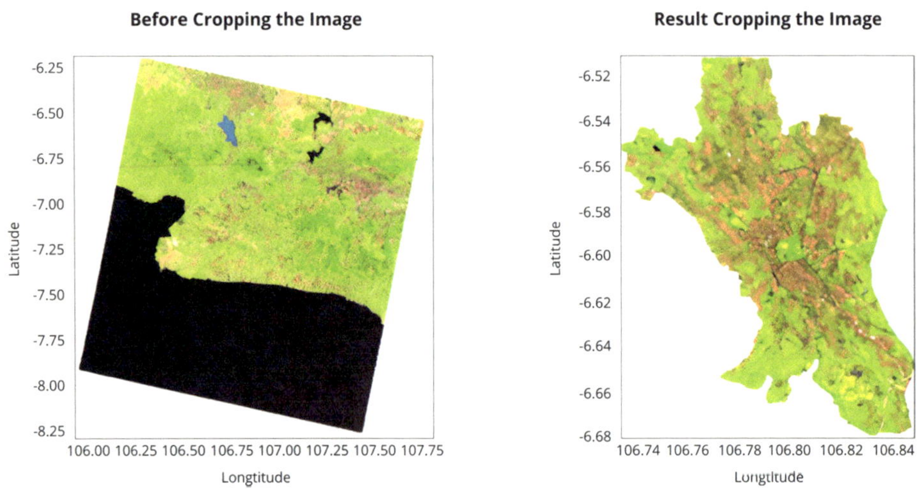


**Figure 4.**
*Image overlay and cropping process.*

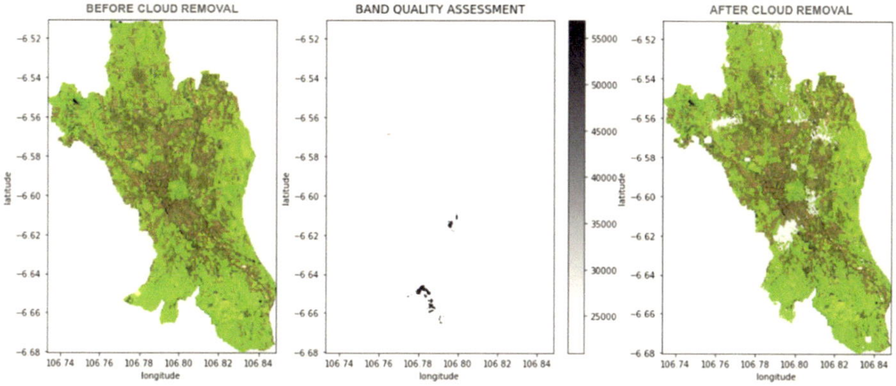


**Figure 5.**
*Cloud removal process and cloud shadows.*

must be processed. This reduction in data processing is hundreds of times faster without sacrificing spatial resolution or analysis quality.

This approach highlights the importance of pre-processing steps, such as administrative boundary overlay and image cropping, in geospatial analysis. These steps ensure that computing resources are focused on the area being studied, thereby increasing efficiency and accuracy. The cropped imagery provides a clearer picture of vegetation and land use conditions in Bogor City, making it more practical for urban planning, environmental monitoring, and other local applications.

**Figure 5** shows the process of removing clouds and shadows on satellite imagery to ensure more accurate vegetation analysis. This process uses Band Quality Assessment (BQA) to identify cloud pixels and shadows, where clouds are shown in dark gray while shadows are shown in black. These elements often interfere with vegetation analysis by distorting the reflectance data.

After the process of cloud and shadow removal, the resulting vegetation analysis image reveals the cleared area with valid pixel values. The white area in the processed image shows the region where the cloud and shadow data have been successfully covered, and the underlying vegetation information is used accurately.

This preprocessing step is essential to improving the reliability of vegetation indices generated from satellite imagery, such as NDVI. By eliminating distortions caused by clouds and their shadows, this analysis provides a more realistic representation of vegetation health conditions and coverage. This process supports accurate monitoring of environmental conditions and is critical for a wide range of applications, including land use planning, agricultural management, and ecological studies.

## 3. Utilization of the internet of things (IoT) in forest monitoring

The internet of things (IoT) has opened up new opportunities for forest monitoring and management. Using a network of interconnected features, IoT can collect real-time area information from various sensors placed in the field [14]. This technology provides practical access to data, which means it is linked to the state of the forest to carry out more efficient and responsive monitoring of changes or threats to forest ecosystems. Forest monitoring with IoT links some of the key components that work in an integrated manner:

1. Area sensor: This sensor is placed in various positions in the forest to measure area parameters such as temperature, humidity, pollution level, soil moisture, and severity of the rays. Data collected by the area sensor can be used to find microclimate changes or early signs of forest fires [5].

2. Sound sensors and cameras: IoT sensors and sound cameras often find human or wild animal activities. Sound sensors can detect engine noise, which could indicate illegal activities such as illegal logging. Cameras connected to IoT networks can also transmit real-time photos and videos, making it easier to monitor forest conditions directly [16].

3. Communication network: IoT features require a communication network to send information to a processing center. Technologies such as LPWAN (Low Power Wide Area Network), GSM networks, and even more so satellites connect sensors in remote locations. This network technology can effectively transmit information from sensors scattered at some point [17].

4. Data processing platform: The information collected by IoT sensors is processed through a platform capable of analyzing and presenting data in an easy-to-understand format. The platform's Artificial Intelligence (AI) technology can help predict events, such as wildfires or changes in rainfall patterns, based on the data collected [18].

Benefits of Using IoT in Forest Monitoring:

1. Early detection of forest fires: IoT sensors can detect fire risks early by monitoring temperature and humidity in real-time. These data can be used to provide early warning and help authorities take preventive measures.

2. Forest ecosystem health monitoring: The health condition of forest ecosystems can be monitored comprehensively through data obtained from humidity, temperature, and air quality sensors. For example, these data can detect environmental stresses on trees due to drought or pollution.

3. Law enforcement against illegal activities: IoT also helps monitor illegal activities within forests, such as illegal logging or illegal hunting. Sound sensors and cameras can detect human presence and suspicious activity and then alert authorities for further action.

4. Wildlife conservation: Sensors installed to monitor wildlife activity provide data on their migration patterns, habits, and habitats. This information can be used to maintain the balance of ecosystems and protect endangered species.

### 3.1 IoT sensors for environmental quality monitoring

IoT sensors have an important role in monitoring the quality of the forest environment by measuring various related variables, such as temperature, air humidity, soil moisture, and air quality. These sensors are strategically placed in the field, and continuous and real-time monitoring is very useful for tracking changes in environmental conditions and detecting early signs of potential threats to forest ecosystems.

1. Temperature sensor

   Temperature sensors measure the air and soil temperature in forests, which are important indicators of environmental conditions. Dramatically rising temperatures, especially during the dry season, can indicate the risk of wildfires. In addition, temperature fluctuations provide useful data for studying microclimate change and its effects on plant and animal life in forests [19].

2. Air humidity sensor

   Air humidity sensors measure the level of humidity in an area within a forest. This level of humidity can vary depending on the season and weather changes. The data generated are very useful for monitoring forest fire risk and plant health conditions. For example, very low humidity can increase the risk of wildfires, while high humidity can indicate areas that favor the growth of certain types of crops [19].

3. Soil moisture sensor

   Soil moisture sensors measure the moisture content in the soil, which is important for assessing the availability of water for forest vegetation. These data are useful for monitoring plant health conditions and detecting early signs of drought or climate change. Soil moisture also plays an important role in the study of forest ecosystems because it is closely related to biological activities, such as the decomposition of organic matter and the distribution of nutrients that affect soil fertility [20].

4. Air quality sensor

   Air quality sensors in forests can measure concentrations of pollutants such as carbon dioxide (CO2), carbon monoxide (CO), ozone (O3), sulfur dioxide (SO2), and air particles. High concentrations of pollutants can come from human activities around forests or from forest fires. Air quality sensors also detect changes in air composition due to seasonal changes or other environmental phenomena. These data are important in assessing ecological health and identifying threats to flora and fauna within forests [21].

These sensors in IoT networks can perform continuous data collection and real-time information transmission to monitoring centers. The advantages include the following:

1. Early detection of fire risk: A combination of temperature, air humidity, and soil moisture data can detect conditions that increase the risk of wildfires. With this data, preventive measures can be taken immediately to reduce potential losses.

2. Monitoring climate change impacts: Long-term data obtained from soil temperature and humidity sensors can aid in the study of microclimate changes in forests. This information provides insights into how global climate change is affecting local ecosystems.

3. Effective natural resource management: Soil moisture and air quality data can help forest managers make decisions based on accurate data, for example, in determining conservation or reforestation strategies appropriate to environmental conditions.

### 3.2 IoT integration with geographic information systems (GIS)

Integrating the internet of things (IoT) with geographic information systems (GIS) improves area monitoring and analysis expertise. By mixing information collected by IoT sensors in the field with GIS platforms, the data obtained can be visualized in real-time and analyzed spatially. This is particularly useful in forest monitoring, where users can view the geographical distribution of area conditions as well as analyze the changes that occur in an area in more detail [22].

1. Real-time data collection from IoT sensors

   Field-installed IoT sensors continuously collect environmental data, such as air temperature, air humidity, soil moisture, and air quality. This data is typically sent to a central server via a communication network such as LPWAN, GSM, or satellite. On those servers, the raw data is stored and ready for further analysis. This integration can help provide up-to-date information to monitor environmental conditions in real-time [23].

2. Real-time visualization on GIS platforms

   The data collected by IoT sensors in the field can be directly uploaded to the GIS platform, where the information is visualized in the form of interactive maps. The GIS platform shows information in layers, sharing data on each area variable, such as soil moisture temperature and air quality, to be analyzed separately or coinciding with a digital map. This visualization looks at the distribution of area conditions and recognizes certain patterns, such as areas with high temperatures that are prone to fires or zones with low humidity. For example, a forest zone faces a significant increase in temperature or shrinkage in moisture. In this regard, GIS maps can share visual warnings, helping stakeholders to quickly follow up and avoid further risks such as forest fires or area degradation [16].

3. Spatial data analysis for decision making

   The integration of IoT data with GIS is not only for visualization but also supports deeper spatial analysis. This analysis is useful in assessing environmental impacts and changes in forest ecosystems [24].

IoT (internet of things) devices and geographic information systems (GIS) complement each other in many applications to collect, analyze, and visualize location-based data. Integration of IoT devices with the GIS of various sectors, such as transportation, agriculture, health, and city management, to efficiently manage spatial data (**Figure 6**) [25].

1. NodeMCU V3 CH340

   NodeMCU V3 CH340 is a microcontroller board based on ESP8266, a WiFi module that can connect to the internet network. The NodeMCU V3 has a CH340 chip as a USB-to-serial converter, making it easier to program and communicate with computers via USB. These boards are widely used in IoT projects because they support wireless connections and are equipped with GPIO (General Purpose Input Output) to connect various sensors and modules. NodeMCU V3 is

**Figure 6.**
*Sensors and materials.*

ideal for projects such as home automation, environmental monitoring, and remote data-driven applications.

2. GPS module UBLOX NEO-6 M-V2

The UBLOX NEO-6 M-V2 GPS module is a GPS signal receiver used to determine geographic location accurately. It has a fast antenna and satellite signal search system and obtains real-time location data such as latitude, longitude, and speed. The UBLOX Neo-6 M-V2 GPS is often used in IoT projects that require location information, such as vehicle tracking, navigation, or other applications that require geographic data.

3. Sensor DHT11

The DHT11 sensor is an easy-to-use and popular temperature and humidity sensor for IoT projects. It can measure temperature in the range of 0–50°C and humidity in the range of 20–90% RH with sufficient accuracy for non-critical applications. These sensors transmit data in digital form, making it easy to connect with microcontrollers such as NodeMCUs. DHT11 is often used to monitor environmental conditions in smart home projects, smart farms, or weather monitoring devices.

4. Cable jumper

Jumper cables are small cables used to connect electronic components on a breadboard or to connect sensors and modules with microcontrollers. These cables come in different types, such as male-to-male, male-to-female, or female-to-female, which can perform connection flexibility. Jumper cables are essential in IoT projects because of the quick and easy connection between various components, especially at the prototype stage.

**Figure 7.**
*Sensor installation.*

**Figure 7** shows the installation of a series of sensors consisting of UBLOX NEO-6 M GPS material, DHT11 temperature and humidity sensors, and a NodeMCU V3 CH340 microcontroller. These components are connected via a breadboard using jumper cables to facilitate connection and experimentation. The CH340 V3 NodeMCU is a control center that wants to receive position information from GPS material as well as temperature and humidity information from the DHT11 sensor.

**Figure 8.**
*Field survey process and placement of environmental monitoring sensors in forest areas.*

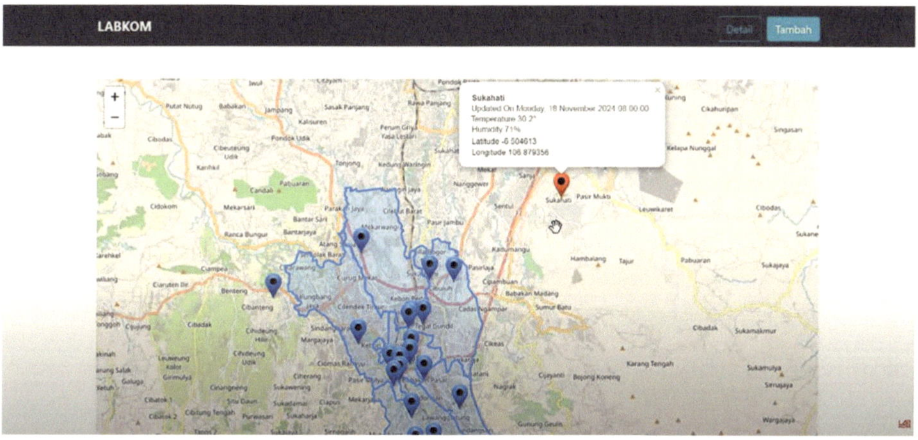


**Figure 9.**
*Digital map-based environmental monitoring website display for forest sensor data.*

The information collected by the sensor can then be sent via a NodeMCU WiFi connection, monitoring the state of the area remotely in real-time.

**Figure 8** explains the location survey activities in forest areas to place environmental monitoring sensors. These sensors are designed to collect environmental data in real-time, including air temperature, humidity levels, and geographical coordinate points in the form of latitude and longitude. These data are also equipped with a record of the date and time of collection so that the information received is accurate and relevant to the current conditions.

**Figure 9** shows the implementation of the monitoring system through an interactive website. This website can access the information collected by the sensor directly. The temperature, humidity, and geographical position of each information collection point can be displayed as an easy-to-understand digital map. With this system, users can effectively monitor the state of the forest and get a reflection of the atmosphere of the area in a certain position without being located in the field.

## 4. Conclusions

Remote sensing technology and the internet of things (IoT) in forest monitoring have opened the way for more accurate, effective, and prolonged monitoring procedures for forest area conditions. Through remote sensing, ecological information can be collected on a wide scale and non-invasively using sensors installed on satellites and aircraft. With accurate results and lightning speed, this technology can monitor various parameters such as vegetation health, land cover change, and illegal activities. Vegetation indices such as NDVI and EVI are meaningful analytical tools that take into account the health of forest ecosystems in an orderly manner.

Meanwhile, IoT technology can carry out real-time monitoring through various sensors placed in the field. These sensors can measure temperature, humidity, air and soil, and air quality. The information produced can be accessed directly and analyzed practically to find threats in early areas, such as fires or forest degradation. Integrating IoT with remote sensing and geographic information systems (GIS) continues to strengthen monitoring expertise with detailed position-based information visualization.

Forest managers, researchers, students, and authorities can benefit from practicing remote sensing technology and IoT in forest monitoring. This technology supports proactive forest conservation and more informed decision-making to conserve forest ecosystems and natural energy sources.

Recommendations include prioritizing investment in advanced sensor networks, capacity-building programs for forest managers, and the integration of artificial intelligence (AI) for predictive analysis. Encouraging international collaboration and open data sharing can further enhance the scalability and effectiveness of forest monitoring systems globally.

## Acknowledgements

The author would like to thank the University of Pakuan for helping and participating with the research team in Bogor, Indonesia, which provided facilities, infrastructure, and resources for this research. Furthermore, the author would like to thank all parties involved, both individually and organizationally, either directly or indirectly, for supporting the materials for the preparation of this book chapter from the research data that has been carried out.

## Conflict of interest

The authors declare no conflict of interest.

## Notes/thanks/other declarations

The author would like to express gratitude to all collaborators and contributors who have provided technical and academic assistance in the research process and book preparation, as well as access to research facilities and resource support.

## Author details

Asep Denih
University of Pakuan, Bogor, West Java, Indonesia

*Address all correspondence to: asep.denih@unpak.ac.id

## References

[1] Goswami S, Gosmawi R, Verma G. Artificial intelligence-based internet of things for industry 5.0. In: Kalloniatis C, editor. The Role of Cybersecurity in the Industry 50 Era. London, UK: IntechOpen; 2024. Available from: https://www.intechopen.com/online-first/89614

[2] Selmy SAH, Kuche DE, Yang Y. Geospatial data: Acquisition, applications and challenges. In: Exploring Remote Sensing - Methods and Applications. London, UK: IntechOpen; 2024. Available from: https://www.intechopen.com/online-first/1192070

[3] Espíndola RP, Ebecken NFF. Advances in remote sensing for sustainable forest management: Monitoring and protecting natural resources. Revista Caribeña de Ciencias Sociales. 2023;**12**(4):1605-1617. Available from: https://revistacaribena.com/ojs/index.php/rccs/article/view/2924

[4] Mahjoobi E, Rafiei F. Exploring the potential of geospatial data: An In-depth investigation. In: Abdalla RM, editor. Exploring Remote Sensing - Methods and Applications. London, UK: IntechOpen; 2024. Available from: https://www.intechopen.com/online-first/1193409

[5] Lechner AM, Foody GM, Boyd DS. Applications in remote sensing to forest ecology and management. One Earth. 2020;**2**(5):405-412. Available from: https://linkinghub.elsevier.com/retrieve/pii/S2590332220302062

[6] Gao Y, Skutsch M, Paneque-Gálvez J, Ghilardi A. Remote sensing of forest degradation: A review. Environmental Research Letters. 2020;**15**(10):103001. Available from: https://iopscience.iop.org/article/10.1088/1748-9326/abaad7

[7] Torresan C, Benito Garzón M, O'Grady M, Robson TM, Picchi G, Panzacchi P, et al. A new generation of sensors and monitoring tools to support climate-smart forestry practices. Canadian Journal of Forest Research. 2021;**51**(12):1751-1765. Available from: https://cdnsciencepub.com/doi/10.1139/cjfr-2020-0295

[8] Kanga S. Advancements in remote sensing tools for forestry analysis. Sustainable Forestry [Internet]. 13 Oct 2023;**6**(1):1-24. Available from: https://systems.enpress-publisher.com/index.php/SF/article/view/2269

[9] Kulkarni R, Honda K. Estimating rice LAI using NDVI: A method for plant conservation education. In: Vegetation Index and Dynamics - Methodologies for Teaching Plant Diversity and Conservation Status. London, UK: IntechOpen; 2023. Available from: https://www.intechopen.com/chapters/1155592

[10] Dutta D, Balha A, Kumar Das P, Jain P, Lukose L, Akram W. Assessment of ecological disturbance caused by flood and fire in Assam forests, India, using MODIS time series data of 2001-2011. In: Remote Sensing. London, UK: IntechOpen; 2021. Available from: https://www.intechopen.com/books/remote-sensing/assessment-of-ecological-disturbance-caused-by-flood-and-fire-in-assam-forests-india-using-modis-tim

[11] Dainelli R, Toscano P, Di Gennaro SF, Matese A. Recent advances in unmanned aerial vehicle forest remote sensing—A systematic review. Part I: A general framework. Forests. 2021;**12**(3):327. Available from: https://www.mdpi.com/1999-4907/12/3/327

[12] Diccion AJL, Duran JZ. Modeling and forecasting cumulative Evi anomalies using sarima for biophysical monitoring: A case study In the Philippines. The International Archives of the Photogrammetry, Remote Sensing and Spatial Information Sciences. 2021;**XLVI-4/W6**:141-146. Available from: https://isprs-archives.copernicus.org/articles/XLVI-4-W6-2021/141/2021/

[13] Venter ZS, Scott SL, Desmet PG, Hoffman MT. Application of landsat-derived vegetation trends over South Africa: Potential for monitoring land degradation and restoration. Ecological Indicators. 2020;**113**:106206. Available from: https://linkinghub.elsevier.com/retrieve/pii/S1470160X20301436

[14] Denih A. In: Kurnia E, editor. Estimasi Evapotranspirasi Menggunakan Teknologi Pengindraan Jauh di Kota Bogor. Sleman: Komojoyo Press; 2023. Available from: https://komojoyopress.com/2023/04/10/estimasi-evapotranspirasi/

[15] Trevisiol F, Mattivi P, Mandanici E, Bitelli G. Cross-sensors comparison of popular vegetation indexes from landsat TM, ETM=, OLI, and sentinel MSI for time-series analysis over Europe. IEEE Transactions on Geoscience and Remote Sensing. 2024;**62**:1-16. Available from: https://ieeexplore.ieee.org/document/10360186/

[16] Raju PS, Lakshmi Priya S. Green IoT framework for deep forest surveillance. In: Proceedings of 2023 International Conference on Signal Processing. London, UK: IEEE; 2023. pp. 1-5. Available from: https://ieeexplore.ieee.org/document/10170016/

[17] Ahmad S, Uyanık H, Ovatman T, Sandıkkaya MT, De Maio V, Brandić I, et al. Sustainable environmental monitoring via energy and information efficient multinode placement. IEEE Internet of Things Journal. 2023;**10**(24):22065-22079 Available from: https://ieeexplore.ieee.org/document/10210671/

[18] Krishnamoorthy M, Asif M, Kumar PP, Nuvvula RSS, Khan B, Colak I. A design and development of the smart forest alert monitoring system using IoT. Journal of Sensors [Internet]. 8 Jan 2023;**2023**(1):1-12. Available from: https://onlinelibrary.wiley.com/doi/10.1155/2023/8063524

[19] Rodrigues CC, da SKP, W de B P, ATM B, de AJFS, PS da S C, et al. IoT sensory platform for monitoring temperature, relative humidity of the air and CO2 in situ. Revista Ibero-Americana de Ciências Ambientais. 2022;**13**(2):278-291. Available from: https://www.sustenere.co/index.php/rica/article/view/7062

[20] Coelho AD, Dias BG, de Oliveira Assis W, de Almeida Martins F, Pires RC. Monitoring of soil moisture and atmospheric sensors with internet of things (IoT) applied in precision agriculture. In: 2020 XIV Technologies Applied to Electronics Teaching Conference (TAEE). New York, USA: IEEE; 2020. pp. 1-8. Available from: https://ieeexplore.ieee.org/document/9163766/

[21] Mittal U, Pawar A, Varshney G, Yadav Y, Sharma R, Satyajeet. IoT based smart monitoring of environmental parameters and air quality index. In: 2021 IEEE 8th Uttar Pradesh Section International Conference on Electrical, Electronics and Computer Engineering (UPCON). New York, USA: IEEE; 2021. pp. 1-5. Available from: https://ieeexplore.ieee.org/ document/9667665/

[22] Safari Bazargani J, Sadeghi-Niaraki A, Choi SM. A survey of GIS

and IoT integration: Applications and architecture. Applied Sciences. 2021;**11**(21):10365. Available from: https://www.mdpi.com/2076-3417/11/21/10365

[23] Cui F. Deployment and integration of smart sensors with IoT devices detecting fire disasters in huge forest environment. Computer Communications. 2020;**150**:818-827. Available from: https://linkinghub.elsevier.com/retrieve/pii/S014036641931360X

[24] Chen L, Mao Y, Zhao R. GIS application in environmental monitoring and risk assessment. In: 2022 3rd International Conference on Geology, Mapping and Remote Sensing (ICGMRS). New York, USA: IEEE; 2022. pp. 908-917. Available from: https://ieeexplore.ieee.org/document/9849269/

[25] Denih A, Kurnia E. In: Kurnia E, editor. Sistem informasi geografis terintegrasi dengan internet of things (IoT) serta penerapan studi kasus. Sleman: Komojoyo Press; 2022

Chapter 5

# Perspective Chapter: FSC Forest Certification and New Technologies as Allies of Sustainable Forest Management in Mexico

*Emily García Montiel, Pablito Marcelo López Serrano, Eusebio Montiel Antuna, Jesús Alejandro Soto Cervantes and Alma Angelina Hernández Rodríguez*

## Abstract

A primary objective of forest certification is to ensure that forest management operations and administration provide opportunities for social, economic and environmental development. This mechanism was created to achieve forest sustainability through good forest management practices. It also helps to maintain transparent and accountable processes in forestry companies, as well as defined processes to promote the quality of operations. The emergence of new geospatial technologies derived from different remote sensing platforms is transforming forest monitoring and management in different ways. These technologies not only improve the efficiency of forest monitoring but also generate information that can contribute to the conservation and sustainable management of forest resources, so it is expected that the integration of information derived from these technologies, such as aerial forest biomass, as an indicator of forest sustainability and FSC environmental criteria and indicators assessment tools will facilitate certification through the detection of patterns and prediction of changes in forest ecosystems. This chapter highlights the importance of the global trend of continuous improvement in forest management processes, certification, and the incursion into new technologies, which together could represent important differences for sustainable forest management in Mexico.

**Keywords:** forest certification, new technologies, sustainability, impacts, Mexico

## 1. Introduction

The importance of forests for the economy, society and biodiversity protection requires that they be managed in a sustainable manner to preserve all their functions [1]. In Mexico, the relevance of forests is reflected in the environmental, economic and social benefits they provide. The forest ecosystem in Mexico provides several vital services that are important for sustainability, such as support services like soil formation, primary production, nutrient cycling and biological control, other services include provision services, such as water, food, raw materials for different sectors, renewable resources, genetic, medicinal and ornamental resources. It also provides regulation services by regulating the climate and air quality and by reducing damage from natural disasters. In addition, cultural services such as scenic beauty, science, education and recreation, all these services are strategic for collective social and economic development in harmony with the environment.

Mexico is one of the 12 megadiverse countries in the world and is home to 70% of the world's flora and fauna species; it has 138.7 million hectares of forest area, which represents 70.6% of Mexican territory [2]. The importance of this sector for climate change mitigation and adaptation is reflected in the capture of 188 million tons of carbon dioxide ($CO_2$). In Mexico, 11.87 million people live in forest areas, of which 3.6 million are indigenous peoples. This population conforms to a modality of social land tenure in our country, classified into ejidos and communities, which are structured in agrarian groups. By 2021, according to [2], timber forest production amounted to 9.3 million $m^3$ of roundwood, of which 8.1 million $m^3$ came from native forests and only 1.2 million $m^3$ from commercial plantations. According to the National Forestry Information System [3], in the last decade, the value of timber and non-timber production reached an average value of 71,021.15 million pesos and the forestry sector generated 266,336 direct jobs. However, production levels remain below the existing potential for harvesting, processing and marketing.

These ecosystems include coniferous forests, broadleaf forests, mountain mesophyll forests, evergreen and sub-evergreen forests and deciduous and sub-deciduous forests. These ecosystems constitute an essential element for the development of human life. In Mexico, 66.7 million hectares are wooded forest areas, representing 48.1%, mainly covered by forests and jungles. Of these, 34.8 million hectares correspond to the forest area for forest ecosystems, representing 25.1% (**Table 1**, **Figure 1**) [2].

Under this scenario, the forestry sector in Mexico faces different environmental problems such as deforestation, according to the National Forestry Commission [2], in the last two decades, approximately 207,665 hectares have been lost annually, in addition to social problems such as population pressure and migration, poverty and marginalization, rural security and economic problems associated with employment. The deficient

| Ecosystem | Forest area in ha | Percentage (%) |
|---|---|---|
| Forests | 34,846,607 | 25.1 |
| Rainforests | 30,332,322 | 21.9 |
| Mangrove | 947,893 | 0.7 |
| Other associations | 527,054 | 0.4 |
| Xerophytic scrub | 56,200,206 | 40.5 |
| Other forest areas | 15,841,147 | 11.4 |
| Total | 138,695,230 | 100 |

*Source: Own elaboration based on CONAFOR, 2024.*

**Table 1.**
*Distribution of forest area in Mexico.*

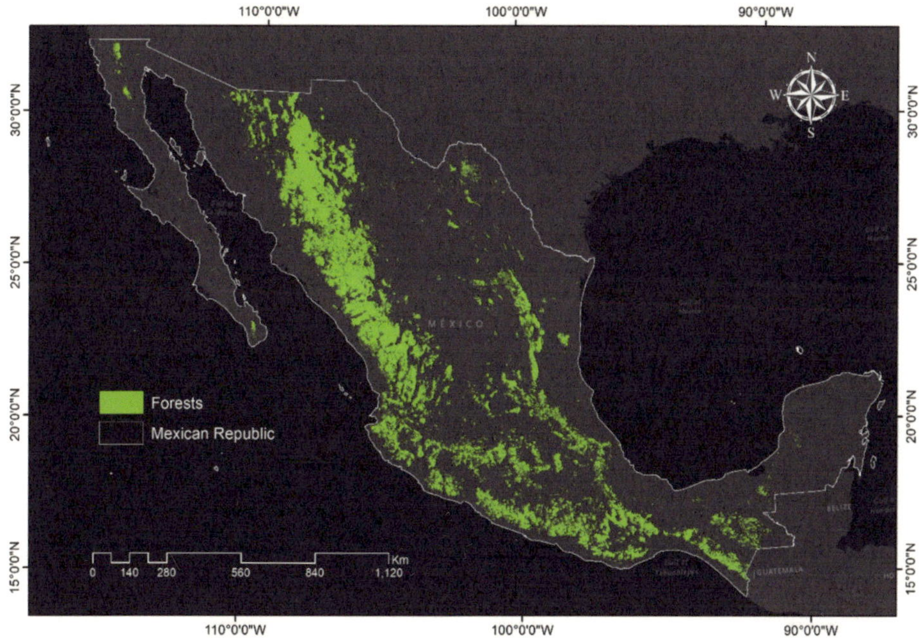


**Figure 1.**
*Spatial distribution of forest ecosystems in Mexico.*

design and development of public policies for rural areas, lack of technological development, governance, social organization, development of social capital and climate change, among other factors, can explain the accelerated process of deterioration of forest ecosystems. In view of these challenges and as a global strategy to promote the sustainable development of forests, forest certification was born in the 1990s as a voluntary mechanism that promotes proper management from the environmental, social and economic point of view [4]. According to Zubizarreta et al. [5], forest certification is an independent third-party assessment system that determines whether forest management meets ecological, economic and social standards. The certification aims to guarantee that forests are sustainably managed based on standards to assure consumers that the products they purchase come from properly managed forests [6].

In the world, there are different forest certification systems, such as the Forest Stewardship Council (FSC), the Lembaga Ekolabel Indonesia (LEI), the Sustainable Forest Initiative (SFI) of the United States and the Programme for the Endorsement of Forest Certification Schemes (PEFC), [7] however, two of the most recognized and relevant are the Programme for the Endorsement of Forest Certification Schemes (PEFC) and the Forest Stewardship Council (FSC) [8]. The PEFC has 296,914,877 million hectares certified worldwide and 13,010 chain of custody certificates in 43 countries [9], while the FSC has 160,446,739 million hectares certified and 63,784 chain of custody certificates in 89 countries, including Mexico [10].

Mexico was one of the first countries to adopt Forest Stewardship Council (FSC) certification in 1993 and was the global headquarters of the FSC from 1993 to 2003. To date, it is the only international certification system operating in Mexico. FSC has ten international principles that all forestry operations must adhere to before they can receive FSC forest management certification. These principles cover a wide range of issues, from maintaining high conservation values to community relations and labor rights, as well as monitoring the environmental and social impacts of forest management. These principles were designed to be applicable worldwide and relevant to all types of forest ecosystems, as well as to a wide range of cultural, political and legal settings [10].

Throughout the more than 30 years since the emergence of this certifying body, different certification modalities have been developed to cover all the links in the forestry sector's production chain. This is the case in Mexico, where there is (a) FSC forest management certification, which confirms that a specific area of the forest is being managed in accordance with FSC principles and criteria. On the other hand, (b) the chain of custody guarantees that products from FSC-certified forests are separated and identified from other non-certified materials throughout the entire processing, manufacturing and distribution chain. Also, (c) ecosystem services are the benefits that people obtain from nature. Forests provide society with a wide range of benefits, from reliable flows of clean water to productive soils and carbon sequestration. In FSC-certified forests, valuable ecosystem services are protected, in 2018, the FSC introduced a procedure to demonstrate, ensure and communicate the positive impact of responsible forest management on ecosystem services. These verified positive impacts aim to facilitate payments for ecosystem services and provide access to other benefits, thus adding commercial value.

Another form of certification arises as retailers see added value in promoting the fact that a product or its packaging is FSC certified. This is where FSC's (d) promotional license comes into play. This means that if a company buys finished products with the FSC label from an FSC-certified company and sells them to end users, it can apply for a promotional license from FSC. These licenses allow companies to use the FSC trademarks to promote these products without the need for certification. Likewise, organizations that use, but do not sell, FSC-certified products within their business can apply for a promotional license.

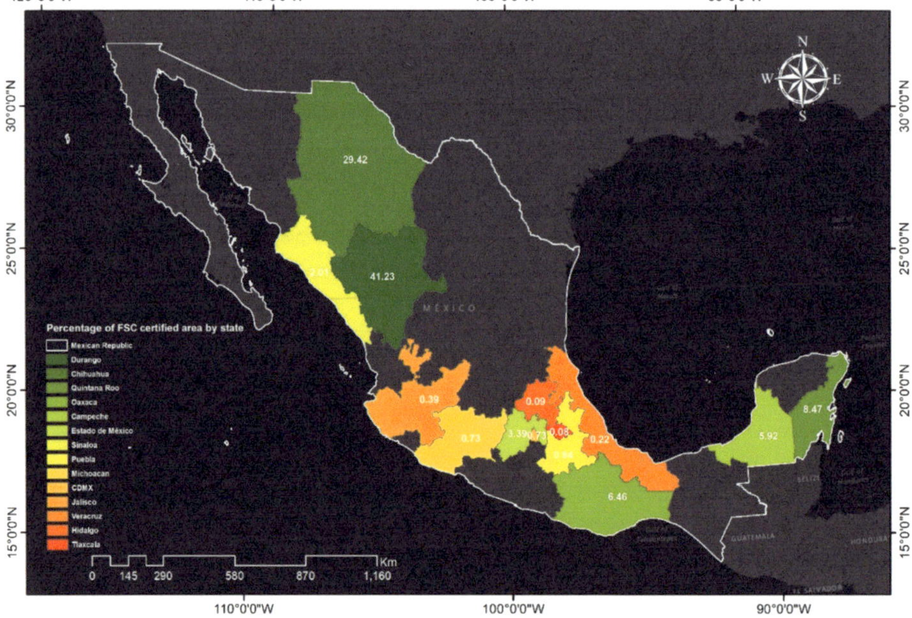


**Figure 2.**
*Distribution of FSC-certified forest area in Mexico.*

| State | Certified area (Has) | Percentage |
|---|---|---|
| Durango | 617,392.80 | 41.23 |
| Chihuahua | 440,611.09 | 29.42 |
| Quintana Roo | 126,874.59 | 8.47 |
| Oaxaca | 96,800.90 | 6.46 |
| Campeche | 88,614.40 | 5.92 |
| Estado de México | 50,808.89 | 3.39 |
| Sinaloa | 30,082.78 | 2.01 |
| Puebla | 12,558.62 | 0.84 |
| Michoacán | 10,967.82 | 0.73 |
| CDMX | 10,960.03 | 0.73 |
| Jalisco | 5902.03 | 0.39 |
| Veracruz | 3273.04 | 0.22 |
| Hidalgo | 1372.04 | 0.09 |
| Tlaxcala | 1205.60 | 0.08 |
| Total | 1,497,424.63 | 100 |

**Table 2.**
*Distribution of certified forest hectares in Mexico by state.*

Currently, Mexico has 1.50 million certified hectares distributed across fourteen states (**Figure 2**). Durango, Chihuahua, Quintana Roo, Oaxaca and Campeche are the states with the highest number of certified forest hectares, standing out from the others by a considerable difference (**Table 2**), distributed in 106 forest management certificates, in addition to 284 chain of custody certificates and 23 promotional licenses for the use of the FSC trademark [11]. Some studies have been carried out to determine the impact of forest certification on the three elements of sustainability, the social, environmental and economic aspects. Although some studies do not report a relevant development in certain indicators that make up forest certification, there are others that show the value that forest properties acquire by having this international seal. One of the main reasons why organizations voluntarily seek forest certification is that through forest certification, they are able to guarantee better prices for their products and entry into new markets. In recent years, the number of certified forest hectares has been increasing, raising the question of whether certification is effective for forest properties and to what extent it has achieved the main purpose for which it was created [12]. Effectiveness in terms of forest certification refers to its ability to reduce deforestation and degradation while maintaining economic viability for forest owners and managers [4].

## 2. Impacts of forest certification

During the period 2010–2020, the world has lost approximately 4.7 million hectares of natural forest area [13], as well as biodiversity at an accelerated rate [14] due to different pressures such as the expansion of agriculture, livestock, forestry and other anthropogenic activities [15, 16]. Loss of biodiversity and climate change are some of the environmental and socioeconomic repercussions that have caused phenomena such as deforestation [17]. On the other hand, forestry regulations are periodically revised and so is FSC certification. However, given this update, the impacts of forest certification on the social, economic and environmental conditions of forests can sometimes not be concluded [18]. Nevertheless, in the face of this environmental crisis, forest certification is a tool that has a positive impact on biodiversity conservation. Throughout more than a decade of forest certification, studies have found positive impacts in different aspects.

For example, Matias et al. [7] report that vascular plant communities are more abundant and richer in FSC-certified forest properties than in non-certified properties, also revealed as another biodiversity indicator that mammals are more abundant in certified properties than in non-certified properties, this may be due to the specifications contained in the different FSC criteria and indicators, such as the requirement to maintain a minimum percentage of the property for exclusive use of biodiversity conservation and maintain the focus on high conservation values [11]. According to Yamamoto et al. [17], forest certification has the potential to improve sustainable forest management, which is its main objective [19]. They report that FSC, in general, produces positive environmental impacts compared to non-certified properties. Specifically, in temperate forests, there are positive impacts on flora, fauna and ecosystem services, highlighting the preservation of species and increasing forest biodiversity in terms of tree richness and density [20].

On the other hand, biodiversity contributes to the generation of ecosystem services, which could be defined as direct and indirect benefits that forests provide to people's wellbeing [21], including tangible goods such as timber, other timber and non-timber products, carbon sequestration, soil retention, cultural, recreational and spiritual values [22]. The results of a study conducted by Paluš et al. [23] show that forest certification contributes positively to the provision of ecosystem services. According to this study, forest owners consider that the greatest impact of forest certification is obtained in biomass provision, water production, erosion control regulation and other natural risks. Additionally, they state that certification helps to promote cultural values of ecosystems such as esthetic, scientific and educational values.

Forest certification has become a strategic tool for companies, specifically to access environmentally sensitive markets and comply with sustainability commitments, [24]. Some of the motivations in economic and social terms for achieving certification are reaching new markets and increasing sales volume and profit margin. In addition to these tangible reasons, there are other intangible reasons, such as gaining market credibility, improving organizational image and upholding ethical and moral values, [5, 25]. According to Dick et al. [26], although there is no clear consensus on the relationship between forest certification and good business performance in financial and economic terms, given that there is the possibility that these efficiency results existed prior to certification, attributed to good business practices, or even to the prior adoption of other quality management systems such as ISO 9001, ISO 14001 or ISO 50001. However, some studies conclude that, although it is complex to attribute good business performance to certification, the forest companies with the best practices and financial efficiency are those that have adopted forest certification [27].

The economic goals of forest certification lie in accessing the timber market in developed countries and ensuring that exported products achieve better prices. Certification promotes forest companies to raise their ecological quality, which in turn provides a competitive effect on exports [28]. Forest certification also plays an important role in the global trade of forest products, with one of its objectives being to create barriers for producers who use illegal timber, contributing to deforestation, forest degradation and harm to indigenous peoples [29].

Forest certification regulates global trade in products and is a tool to ensure that trade in forest products does not compromise the sustainability of these resources [30]. In addition, certification is sometimes a requirement for entering international markets [31]. On the other hand, it is considered an element that brings improvements in the internal processes of forestry organizations and promotes good relations between companies and their partners (suppliers, customers and stakeholders) [8, 32, 33].

A well-managed, economically viable and orderly forest should impact social benefits for the community. Thus, according to Eleine Juliana Malek and Abd Rahman Abdul Rahim [8], one of the motivations for seeking certification is to ensure social development. According to Tricallotis [34], some of the social problems that existed in Chile (similar conditions to Mexico) when forest certification emerged were poor to modest labor conditions for workers, poor training, poor performance in occupational health and safety and overwork, among others, as well as conflicts with indigenous communities over land tenure and few benefits for society. FSC's principles 2–5 seek to guarantee that the community maintains its rights and enjoys the benefits of forest resources for its optimal development [35].

Zubizarreta et al. [31] find a series of external and internal motivations that different studies have addressed, which are the motivations considered by owners in the different countries where forest certification exists. Depending on their conditions and circumstances, the different forest companies may consider them as drivers to seek and maintain certification (**Table 3**).

We can add to the above findings of García-Montiel et al. [36] where a group of 30 experts were consulted as decision-makers to determine which principles, criteria and indicators are considered the most relevant while implementing forest certification. For decision-makers, the environmental principle occupied the first place with 40.26% of importance in the implementation of forest certification, followed by the social principle and the economic principle with 32.15 and 27.59% of importance, respectively. Regarding the criteria, forest management and production, biodiversity and forest protection were considered to be the most relevant. Regarding the indicators, the results indicated that forest certification in Mexico can have a positive impact on the existence of educational institutions and community services such as water, energy, medical services and drainage. It also influences the quality of forest management plans, investment in forest management, machinery and equipment,

| Internal factors | External factors |
|---|---|
| Moral (values of forestry companies) | Market motivation |
| Continuous learning | Prestige |
| Improved economic and operational results | Compliance with regulations |
| Improved performance | |
| Economic benefits | |
| *Source: Own elaboration based on Zubizarreta et al. [31].* | |

**Table 3.**
*Internal and external motivations for forest certification.*

environmental services, recreation, tourism, research, development and community education, as well as planning for the conservation of biological diversity.

These motivations could be used as a testimony so that forest properties that are not yet interested in certification take them into account in their decision to join the recognition of their good practices or see certification as a method for achieving improvements and benefits. According to Garcia et al. [37], forest monitoring, forest protection, chemical use reduction and biodiversity conservation are environmental aspects where certified properties perform better than non-certified properties. Similarly, other planning practices, such as forest management programs, best management practices, care of the harvesting area and treatment of endangered species, are also moderately better in certified properties. In this same study, it was found that in the social aspect, certified farms showed significant changes in the existence of agreements that recognize the legal or traditional rights of the communities, workers have better benefits and working conditions, as well as in the existence of norms that regulate the use of community services, and the inhabitants of certified forest ejidos have significantly greater access to secondary and higher education. For the economic aspect, significant changes were found in certified properties in the existence of program budgets, the efficient use of resources, the freedom of workers to organize, the existence of an accounting system and year-round contracts.

## 3. New technologies in forest management

Forest management is a process of planning and executing activities for the administration and use of forests in order to make them sustainable, i.e. environmentally responsible, socially beneficial and economically viable. A permanent tree cover is an indicator of good forest management, since it ensures a continuous supply of quality water, protects soils from erosion, maintains biological diversity, provides medicinal plants, edible fruits and mushrooms and allows recreation and other productive activities.

In Mexico, there is the figure of technical forest service providers or forest managers, and in order to manage forests, they use the basis of silviculture, which guarantees an adequate and optimal use of the resource that currently must have a sustainability approach. Forest managers must properly manage forests to achieve one or more specific management objectives with respect to the continued production of forest products and services, without reducing their inherent values and ensuring future productivity [38]. For this purpose, forest management programs are employed that must contemplate not only sustainable harvesting systems but also the permanent provision of forest ecosystem services. However, forest management of forest ecosystems in Mexico, particularly in a global context marked by the aforementioned factors of climate change and the pressure to meet growing demands for natural resources, presents increasingly complex challenges.

This represents a challenge for forest managers, while also offering a wide range of opportunities for creativity in developing management models and strategies for managing the resources under their responsibility, as well as internalizing the value of the forest in biodiversity conservation, provision of hydrological services and carbon sequestration [39]. Under this scenario, alternatives are urgently needed to adapt new innovative approaches that not only optimize the exploitation of resources but also ensure their conservation for future generations.

Therefore, the study and monitoring of forest ecosystems have entered a process of innovation worldwide with the use of technologies derived from satellite remote sensing and sensors carried on unmanned aerial vehicles (UAVs). In parallel, algorithms and machine learning methods have been developed through artificial intelligence (AI) applied mainly to the processing and collection of data to measure and evaluate forest ecosystems. In this sense, AI promises to be an innovative tool that will help foresters make more efficient decisions in forest management since its potential allows the prediction of natural and anthropogenic disasters, such as forest fires, as well as the optimization of resource management through predictive models. According to Sotomayor [40], these innovations will allow the transformation of traditional forest management practices, making them more accurate and efficient in decision-making to sustainably manage forest resources.

Despite the potential of new integrated technologies in forest management, in several developing countries, such as Mexico, they are still in their initial stages. The main barriers to the inclusion of these technologies lie in the initial investment in equipment and the specialization of human resources, in addition to the fact that the implementation of these technologies depends on public policies that promote their use, as well as on international cooperation to overcome technical and financial limitations. For this reason, different educational and forestry research institutions in Mexico are beginning to include new methodological approaches for the training of new forestry professionals, as well as in the development of forestry research projects and technologies that allow faster and, in many cases, more precise processes in the study of the structure of forest ecosystems. Some isolated efforts demonstrate these technologies to be novel tools, with different multispectral sensors that provide greater detail with very high-resolution images, which allow monitoring the vigor and health of vegetation, as well as identifying changes in forest mass, land use and/or deforestation, fires, pests and illegal logging. In addition, these technologies have acquired great importance for the quantification of resources, through the more accurate estimation of aerial forest biomass, as an important parameter for monitoring and assessing the current state of forests and, therefore, facilitate better decision-making in the planning and use of forest resources. These applications have led to forest certification as one of the alternatives that promotes the accreditation of a forest area with indicators that guarantee sustainability. The coherent integration of forest certification and new technologies promotes the competitiveness of products that have been produced with forest raw materials from sustainably and responsibly managed forests. In addition to being useful tools to assess the condition of certified forest areas and their surroundings through geospatial data in near real time of forest condition, allowing to reduce certification audits and ensure its monitoring that impacts on social, environmental and economic benefits to society.

## 4. Technical services in northern Mexico

Through a personal consultation with the main people responsible for technical services in northern Mexico, who carry out various activities of planning, implementation and forest harvesting in forest management programs, they consider that FSC forest certification has had a positive impact on the different areas of sustainability.

The following are the opinions expressed literally, with respect to their point of view of the main benefits that forest certification has brought in Mexico:

- "The conservation of biodiversity, contributes to the mitigation of climate change, gives added value to products from forests".
- "It has benefited in the opening to a greater number of markets".
- "Support and economic incentives for ejidos and forest communities".
- "Access to other sources of financing with alternative projects to timber harvesting".
- "Improved rights for forest workers".
- "Improvements in regulation, governance, and business model".
- "Wildlife conservation, reduced presence of fires, increased reforestations".
- "Greater care for forest resources and the environment, governance has been strengthened in certified ejidos, greater integration of women into forestry activities".

Service providers also consider that some impacts have yet to be achieved, such as

- "The commitment and collaboration of governments with forest certification".
- "Improve processes for the transformation of raw materials".
- "Greater economic impacts for forest resource owners".
- "Greater social awareness and impact".
- "Higher price of certified products".

Likewise, the main opinions he has regarding whether geospatial technologies support measuring the impacts of forest certification are as follows:

- "Yes, they make it easier to have information in areas that are difficult to access as well as to help planning in the decision-making process".
- "Yes, because satellite images provide a broader view of the forest terrain".
- "It is graphic and quick information".
- "Yes, you have a source of current and high resolution information".
- "Yes, verifying the forest cover and impacts to the ground".
- "Yes, because you can see the cover of the lands that are certified and make comparisons and see if there is a positive impact on these lands. But a field visit is also necessary to measure other aspects of the forest".
- "Yes it is possible given that we can measure the possible losses of cover in the certified ejidos".

## 5. Conclusions

In conclusion, considering the experience of forestry technical service providers, as well as the literature review and the authors' own experience, forest certification is a tool for the controlled use of forest and environmental resources. It is considered important to have an external evaluation that allows maintaining forests in constant development and continuous improvement. In general, certification is an option to combat global warming and provide social development benefits to communities. Although there are benefits that are not yet perceived in all forest properties such as the increase in product prices, the environmental benefits have made a significant difference. There is still the challenge of promoting sufficient interest and information so that forest owners are motivated to achieve certification and that the costs and complexity of constant monitoring by third parties integrated with traditional methods and new technologies are not a limiting factor in adopting this tool. Certification is a vehicle to international markets and adds to compliance with global trends. Beyond being a voluntary process, in the future, the possibility could be explored of making the criteria of good forest management that certification implies mandatory in order to ensure the preservation of the environment, specifically the world's forests.

## Acknowledgements

We extend our thanks to the Consejo Estatal de Ciencia y Tecnología de Durango (COCYTED) for their support in the financing of this academic work.

## Conflict of interest

The authors declare that they have no conflict of interest.

## Author details

Emily García Montiel[1*], Pablito Marcelo López Serrano[2], Eusebio Montiel Antuna[1], Jesús Alejandro Soto Cervantes[1] and Alma Angelina Hernández Rodríguez[3]

1 Faculty of Forestry and Environmental Sciences of the Juarez University of the State of Durango, Mexico

2 Institute of Forestry and Wood Industry of the Juarez University of the State of Durango, Mexico

3 Master's Program in Geomatics Applied to Forestry and Environmental Resources at the Juarez University of the State of Durango, Mexico

*Address all correspondence to: e_garcia@ujed.mx

## References

[1] Orsi F, Ciolli M, Primmer E, Varumo L, Geneletti D. Mapping hotspots and bundles of forest ecosystem services across the European union. Land Use Policy. 2020;**99**:104840. DOI: 10.1016/j.landusepol.2020.104840

[2] Principales indicadores del sector forestal en México. Comisión Nacional Forestal. Data bosques 2024. Available from: https://databosques.cnf.gob.mx/inicio/

[3] Sistema Nacional de Información Forestal. Secretaria de Medio Ambiente y Recursos Forestales. 2024. Available from: https://snif.cnf.gob.mx/

[4] Wolff S, Schweinle J. Effectiveness and economic viability of Forest certification: A systematic review. Forests. 2022;**13**:798. DOI: 10.3390/f13050798

[5] Zubizarreta M, Arana-Landín G, Cuadrado J. Forest certification in Spain: Analysis of certification drivers. Journal of Cleaner Production. 2021;**294**:1. DOI: 10.1016/j.jclepro.2021.126267

[6] Lewis RA, Davis SR. Forest certification, institutional capacity, and learning: An analysis of the impacts of the Malaysian timber certification scheme. Forest Policy and Economics. 2015;**52**:18-26. DOI: 10.1016/j.forpol.2014.12.011

[7] Matias G, Cagnacci F, Rosalino LM. FSC forest certification effects on biodiversity: A global review and meta-analysis. Science of the Total Environment. 2024;**908**:168296. DOI: 10.1016/j.scitotenv.2023.168296

[8] Malek EJ, Rahim ARA. A thematic review of forest certification publications from 2017 to 2021: Analysis of pattern and trends for future studies. Trees, Forests and People. 2022;**10**:100331. ISSN 2666-7193. DOI: 10.1016/j.tfp.2022.100331

[9] PEFC. 2024. Available from: https://www.pefc.org/

[10] FSC. 2024. Available from: https://fsc.org/en

[11] FSC. 2024. Available from: https://mx.fsc.org/mx-es/datos-y-cifras

[12] Mukherjee I, Bali AS. Policy effectiveness and capacity: Two sides of the design coin. Policy Design and Practice. 2019;**2**(2):103-114. DOI: 10.1080/25741292.2019.1632616

[13] FAO, UNEP. The State of the World's Forests 2020, Forests, Biodiversity and People. Rome: Elsevier B.V; 2020. DOI: 10.4060/ca8642en

[14] Johnson CN, Balmford A, Brook BW, Buettel JC, Galetti M, Guangchun L, et al. Biodiversity losses and conservation responses in the Anthropocene. Science. 2017;**80**(356):270-275. DOI: 10.1126/science.aam9317

[15] Tilman D, Clark M, Williams DR, Kimmel K, Polasky S, Packer C. Future threats to biodiversity and pathways to their prevention. Nature. 2017;**546**(7656):73-81. DOI: 10.1038/nature22900

[16] IPCC. Climate Change 2014: Synthesis Report. Contribution of Working Groups I, II and III to the Fifth Assessment Report of the Intergovernmental Panel on Climate Change [Core Writing Team, R.K.

Pachauri and L.a. Meyer (Eds.)]. Geneva, Switzerland: IPCC; 2014 151 pp

[17] Yamamoto Y, Matsumoto KI. The effect of forest certification on conservation and sustainable forest management. Journal of Cleaner Production. 2022;**363**:132374. DOI: 10.1016/j.jclepro.2022.132374

[18] Van der Ven H, Rothacker C, Cashore B. Do eco-labels prevent deforestation? Lessons from non-state market driven governance in the soy, palm oil, and cocoa sectors. Global Environmental Change. 2018;**52**:141-151. DOI: 10.1016/j.gloenvcha.2018.07.002

[19] Di Girolami E, Arts BJM. Environmental Impacts of Forest Certifications. Wageningen, Netherlands: Wageningen University and Research; 2018. Available from: https://edepot.wur.nl/464632

[20] Kalonga SK, Midtgaard F, Klanderud K. Forest certification as a policy option in conserving biodiversity: An empirical study of forest management in Tanzania. Forest Ecology and Management. 2016;**361**:1-12. DOI: 10.1016/j.foreco.2015.10.034

[21] Pascual U, Muradian R, Brander L, Gómez-Baggethun E, Martín-López B, Verma M, et al. The economics of valuing ecosystem services and biodiversity. In: The Economics of Ecosystems and Biodiversity: Ecological and Economic Foundations. Routledge; 2012. pp. 183-256. DOI: 10.4324/9781849775489

[22] Saarikoski H, Jax K, Harrison PA, Primmer E, Barton DN, Mononen L, et al. Exploring operational ecosystem service definitions: The case of boreal forests. Ecosystem Services. 2015;**14**:144-157. DOI: 10.1016/j.ecoser.2015.03.006

[23] Paluš H, Krahulcová M, Parobek J. Assessment of Forest certification as a tool to support Forest ecosystem services. Forests. 2021;**12**:300. DOI: 10.3390/f12030300

[24] Zubizarreta M, Arana-Landín G, Wolff S, Egiluz Z. Assessing the economic impacts of forest certification in Spain: A longitudinal study. Ecological Economics. 2023;**204**:107630. DOI: 10.1016/j.ecolecon.2022.107630

[25] Kolstad I. Why firms should not always maximize profits. Journal of Business Ethics. 2007;**76**:137-145. DOI: 10.1007/s10551-006-9262-7

[26] Dick GP, Heras I, Casadesús M. Shedding light on causation between ISO 9001 and improved business performance. International Journal of Operations and Production Management. 2008;**28**(7):687-708. DOI: 10.1108/01443570810881811

[27] Naveh E, Marcus A. Achieving competitive advantage through implementing a replicable management standard: Installing and using ISO 9000. Journal of Operations Management. 2005;**24**(1):1-26. DOI: 10.1016/j.jom.2005.01.004

[28] Overdevest C, Rickenbach MG. Forest certification and institutional governance: An empirical study of forest stewardship council certificate holders in the United States. Forest Policy and Economics. 2006;**9**(1):93-102. DOI: 10.1016/j.forpol.2005.03.014

[29] Chen J, Wang L, Li L, Magalhães J, Song W, Lu W, et al. Effect of forest certification on international trade in forest products. Forests. 2020;**11**(12):1270. DOI: 10.3390/f11121270

[30] N'Doua BD. The impact of forest management certification on exports

in the wood sector: Evidence from French firm-level data. Journal of Cleaner Production. 2023;**418**:138032. DOI: 10.1016/j.jclepro.2023.138032

[31] Zubizarreta M, Arana-Landín G, Siguenza W, Cuadrado J. Forest certification and its impact on business management and market performance: The key role of motivations. Forest Policy and Economics. 2024;**166**:103266. DOI: 10.1016/j.forpol.2024.103266

[32] Tricallotis M, Gunningham N, Kanowski P. The impacts of forest certification for Chilean forestry businesses. Forest Policy and Economics. 2018;**92**:82-91. DOI: 10.1016/j.forpol.2018.03.007

[33] George AK, Kizha AR, Daigneault A. Is forest certification working on the ground? Forest manager's perspectives from the northeast US. Trees, Forests and People. 2022;**7**:100197. DOI: 10.1016/j.tfp.2022.100197

[34] Tricallotis M. ¿En qué contexto surge la certificación forestal en Chile?: desempeño ambiental, social y económico de empresas no certificadas. Bosque (Valdivia). 2016;**37**(3):613-624. DOI: 10.4067/S0717-92002016000300018

[35] FSC. 2021. Available from: https://fsc.org/es/newscentre/standards/publicacion-del-estandar-nacional-de-manejo-forestal-responsable-fsc-de-mexico

[36] García-Montiel E, Cubbage F, Rojo-Alboreca A, Morones-Esquivel MM, Lujan-Álvarez C, Montiel-Antuna E, et al. Hierarchical analysis of factors determining the impact of Forest certification in Mexico. Forests. 2022;**13**(12):2093. DOI: 10.3390/f13122093

[37] Garcia E, Corral JJ, López PM, Cubbage F, Ramírez H, Cervantes JA, et al. Socioeconomic and environmental impact of forest certification in ejidos and communities of the state of Durango, Mexico. International Forestry Review. 2023;**25**(4):399-409. DOI: 10.1505/146554823838028238

[38] Santiago-García W, Ángeles-Pérez G, Quiñonez-Barraza G, Santos-Posadas HM, Rodríguez-Ortiz G. Avances y perspectivas en la modelación aplicada a la planeación forestal en México. Madera y bosques. 2020;**26**(2):1-16. DOI: 10.21829/myb.2020.2622004

[39] Aguirre-Calderón OA. Manejo forestal en el siglo XXI. Madera y bosques. 2015;**21**(SPE):17-28

[40] Sotomayor JFM. Inteligencia Artificial en el Manejo Forestal: Una Revisión Sistemática Basada en el Marco Metodológico SALSA. Revista Ingeniería e Innovación del Futuro. 2024;**3**(1):38-53. DOI: 10.62465/riif.v3n1.2024.79

Section 3

# Biodiversity and Conservation Strategies

Chapter 6

# Himalayan Serow: A Keystone Species in Crisis – Challenges and Conservation Strategies

*Iyaz Quyoom, Bilal A. Bhat and Wasim Sajad Malik*

## Abstract

Himalayan serow (*Capricornis sumatraensis thar*), a distinctive member of the Bovidae family, occupies a unique ecological niche in the mountainous terrains of the Himalayas. This chapter explores the serow's taxonomy, distribution, morphology, habitat preferences, diet, behaviour, and conservation status, emphasizing the species' evolutionary adaptations and the critical threats it faces. The serow's solitary nature and preference for densely forested, high-altitude environments have rendered it elusive and under-studied, yet it remains a key indicator of the health of Himalayan ecosystems. Despite its adaptability, the serow is increasingly endangered by habitat fragmentation, poaching, and climate change. The present chapter synthesizes current knowledge on the species, highlighting significant gaps in research, particularly its genetic diversity, disease prevalence, and reproductive strategies. It calls for enhanced conservation efforts, including habitat conservation, community engagement, and further ecological studies. The plight of the Himalayan serow underscores the broader challenges of biodiversity conservation in an era of rapid environmental change, serving as a powerful reminder of the intricate interdependencies within ecosystems and the urgent need for holistic, dynamic conservation strategies.

**Keywords:** high-altitude ecosystems, mountain ungulates, ecological adaptation, habitat preferences, habitat fragmentation, anthropogenic threats, conservation strategies

## 1. Introduction

The Goat-Antelopes, a generic name for serow, goral, and takin, form the third division of the Bovidae, which is known as the Rupicaprinae. They are said to hold an intermediate position between goats and antelopes. The lineage of serows is thought to have originated in Central Asia during the middle to late Pliocene, approximately 3.3 to 2.5 million years ago [1–3]. Currently, the majority of serow populations are found in China, Japan, and Southeast Asia [4, 5].

The mainland serow (*Capricornis sumatraensis*), a widely distributed species in China and Southeast Asia, previously had a range extending as far north as Northern

China (39°N). Despite its extensive distribution across more than 11 countries, the mainland serow faces significant challenges, including poaching, habitat degradation, and fragmentation. These factors have impacted the survival and recovery of the species, resulting in isolated and dwindling populations [6, 7]. Due to these threats combined with a lack of precise data on population size and distribution, implementing effective long-term conservation measures is challenging [2, 5, 8].

Himalayan serow (*Capricornis sumatraensis thar*), a subspecies of the mainland serow, belongs to the subfamily Caprinae under the family Bovidae. Serows are small to medium-sized, goat-like, and are the most generalized representatives of the subfamily Caprinae. Therefore, all Caprinae species are believed to have evolved from a serow-like ancestor. The generic name is derived from two Latin words: Capra which means a goat, and cornu, the horn of an animal; hence, *Capricornis* implies the presence of goat-like horns. *C.s. thar* has been assessed as Vulnerable (VU A2; C1) as this subspecies is experiencing significant declines in population size, range size, and habitat over the last decade, and continuing significant declines are expected to occur in the future due to anthropogenic pressure [4].

Himalayan serow, resembling a hybrid of a goat, cow, pig, and donkey, is restricted to the Himalayan forest slopes and is thought to be of oriental origin [9]. The Himalayan serow is a territorial and solitary ungulate that uses dung piles to demarcate its territory [10]. Due to its specialized requirements for dense and undisturbed forests, the Himalayan serow is considered a flagship species. The serow thrives in densely forested, secluded river valleys and steep, grassy slopes, often in close proximity to thickets of oaks and rhododendrons [11]. It prefers altitudes ranging from 1500 to 4000 m asl and is less adaptable to arid conditions. It prefers moist regions and is a solitary mammal [12]. Its preferences for food and habitat support solitary behaviour as well. In harsh winters, serow has been reported to move towards lower altitudes, but seasonal change in its native ranges in the Himalayas has not been seen [4]. Bangladesh, Bhutan, China, India, and Nepal are currently home to this subspecies [13].

Like other mountain ungulates, serow also forms a part of the prey base for carnivores, especially leopards [8]. The Himalayan serow is provided with a physique and suite of adaptations that are finely tuned to the challenging environments in which it lives. However, it faces an array of threats that imperil its survival. The conservation of this remarkable species has become a major concern, not just for the sake of preserving biodiversity but also for understanding the broader challenges of conserving life in one of the most ecologically sensitive regions on earth [2, 14].

This chapter intends to provide a foundation for informed conservation efforts by conducting a comprehensive evaluation of the current state of knowledge about the serow and emphasizing the importance of safeguarding the rich biodiversity that thrives in the Himalayan heights. While the Himalayan serow may be an enigmatic and rarely seen inhabitant of these high mountains, its story holds a powerful lesson about the preservation of earth's most remarkable life forms against the backdrop of an evolving and challenging world. The objectives of the chapter are: (1) to critically review and synthesize existing literature on various ecological aspects of the Himalayan serow and (2) to identify the gaps for potential future research which may help outline holistic management planning for the delicate mountainous environment.

### 1.1 Taxonomy

Initially, based on their physical characteristics, Wilson and Reeder recognized six species in the genus *Capricornis*: Himalayan serow (*Capricornis thar*), Chinese serow (*Capricornis milneedwardsii*) with the subspecies (*C.m. maritimus*), Red serow (*Capricornis rubidus*), Sumatran serow (*Capricornis sumatraensis*), Japanese Serow (*Capricornis crispus*) and Formosan serow (*Capricornis swinhoei*) [15]. Later, Wilson and Mittermeier elevated the Indochinese serow (*C.m. maritimus*) from subspecies to species level (*C. maritimus*) to make it seven serow species under the genus *Capricornis* [16].

Liu *et al*. used mitochondrial Cytb and D loop sequence analysis to determine the species' molecular phylogenetic position and found that Chinese serow (*Capricornis milneedwardsii*), was most closely related to Himalayan serow (*Capricornis thar*) [17]. A study suggests *C. crispus* as the basal species, while *C. thar* is grouped with *C. milneedwardsii*, which is a sister clade of *C. sumatraensis* [18]. However, a recent analysis using the total mitochondrial genome of all taxa suggests that the genus *Capricornis* has only four species: Japanese Serow (*C. crispus*), Formosan Serow (*Capricornis swinhoei*), Red Serow (*C. rubidus*), and *Capricornis sumatraensis* including subspecies, Sumatran serow (*C.s. sumatraensis*), Chinese serow (*C.s. milneedwardsii*), and Himalayan serow (*C.s. thar*) [6].

### 1.2 Distribution and population status

The Himalayan serow, with its oriental origins, is found in northeast India, Tibet (China), east and southeast Bangladesh, the Himalayas (including Bhutan, northern India, Sikkim, and Nepal), and Myanmar, though its existence there is uncertain [12, 19]. This subspecies has a significant, but limited, population in the Himalayas, stretching from Jammu and Kashmir to Nepal, Bhutan, and the easternmost reaches of Arunachal Pradesh. It naturally occurs at low densities throughout most of its range in India, with a typically fragmented distribution [4].

In the Union Territory of Jammu and Kashmir, the presence of this subspecies has been reported through direct and indirect evidence in several protected areas and forest ranges: Dachigam National Park, Brien-Nishta Wildlife Conservation Reserve, Kishtwar National Park, Overu-Aru Wildlife Sanctuary, Tangmarg-Gulmarg valleys [20], and Bani Wildlife Sanctuary and its adjoining areas [21, 22]. Pandey notes its wide distribution along the Sutlej and Beas rivers in Himachal Pradesh [23]. Recent observations have confirmed their occurrence in several districts, including Chamba, Lower Kinnaur, Shimla, Kangra, and Kullu [24], as well as the Spiti Valley [25].

In Uttarakhand, camera trap surveys in Rudraprayag (Kedarnath Forest Division, Kedarnath Wildlife Sanctuary) and Pauri Garhwal (Lansdowne Forest Division) have documented the species presence [26, 27]. While its distribution in other parts of the state is uncertain, it likely inhabits small forest patches between 500 and 3500 m asl. The serow is also widely distributed in Sikkim, with populations identified in the Khangchendzonga National Park and other areas [20, 28, 29]. In West Bengal, it has been reported from the Darjeeling Wildlife Division (Singalila National Park) and Gorumara Wildlife Division (Neora Valley National Park) [27].

In Assam, the serow inhabits the northern banks of the Brahmaputra River, with sightings in Kokrajhar, Sonitpur, Lakhimpur, and Dhemaji districts [30, 31]. It is also found in the lowland woods up to 3500 m in Arunachal Pradesh, with confirmed populations in various districts [4].

In Nepal, the serow is prevalent in many protected areas such as Annapurna Conservation Area, Sagarmatha National Park, Dhorpatan Hunting Reserve, Langtang National Park, Rara National Park, Chitwan National Park, and Bardia National Park [8]. In Bhutan, it is distributed up to an elevation of 3500 m asl in various national parks. In China, it occurs in the forest belt between 2000 and 3000 m asl on the southern slope of Qomolangma along the Nepal border, with Tibet reporting the highest detection through camera trap surveys [32].

There have been no serow counts in Bangladesh, but their numbers are believed to be minimal. Camera trap surveys in the Sangu Reserve from 2011 to 2015 did not report any serow presence [33]. In Bhutan, Wang estimated a serow density of 0.36 individuals/km$^2$ through distance sampling [34]. In Nepal, no population size estimates exist, but Aryal estimated a density of 1.17 individuals/km$^2$ in the Annapurna Conservation Area, a notable decline over the past decades [35].

In India, the subspecies population size is unknown. However, in areas with sufficient habitat, its density has been assessed to be approximately 1.6 individuals/ km$^2$ in Uttarakhand [36]. In the Khangchendzonga Biosphere Reserve, Sikkim, frequent captures were made through camera traps from 2008 to 2010 [28]. In Arunachal Pradesh's Dibang Valley, Nijhawan estimated serow densities ranging from 0.26 to 1.46 individuals/km$^2$ in various areas, including community forests and the Dibang Wildlife Sanctuary [37]. There are no population estimates from Jammu and Kashmir and Himachal Pradesh, and camera trap investigations indicate infrequent sightings, suggesting naturally low numbers even in protected areas [4].

## 2. Morphology

The serow possesses an ungainly build characterized by a large head, short limbs, and a thick neck. Its peculiar stance, with forelegs astride, hoofs widely splayed, and head thrust downward, contributes to its awkward appearance [38]. Both sexes share a similar build. An adult serow's shoulder height ranges between 100 and 110 cm. In terms of size, serows vary from 140 to 180 cm in length and typically weigh around 90 kg. Both males and females have horns that are conical, backwards-curving, and sharply pointed, with lengths averaging between 18 and 25 cm, though reports of horns up to 32 cm exist [12, 38]. Distinguishing features include a large head, long ears akin to those of mules, a thick neck, and short limbs. Their thick coat varies in colour from red to grizzled black, blackish-gray, and roan. The Himalayan serow is characterized by dirty white undersides and reddish-brown limbs above the knee. Sexual dimorphism in this species is minimal. In females, the oestrus state is indicated by the size of their preorbital glands, which become enlarged for scent marking [39].

### 2.1 Habitat preferences of the serow

The mainland serow (*Capricornis sumatraensis)*, a forest-dwelling ungulate, inhabits mountain slopes with rugged, steep hills and rocky places from 100 m to 4000 m asl [40, 41]. According to Prater and Schaller, serows often inhabit gorges that are moist and heavily forested and can be found at elevations of 1500–4000 m asl [12, 38]. Serows have been frequently found in subalpine and temperate habitats in the low and mid-elevation range of 1200–3700 m asl with a preference for higher elevation >2100 m in Sikkim, India [28]. According to a study, serows prefer 2500 to 3500 m asl altitude range in the central Himalayas of Nepal [9]. Higher altitudes (>4000 m asl) are used to flee from predators in Nepal, while elevations between 2500 and 4000 m asl are mostly used for food and shelter. They often avoid plains and prefer areas with gentle to steep hills. Steeper slopes are used as resting places, while gentle slopes are preferred for grazing. To protect themselves from inclement weather, predators, and hunting, they make use of varied cover (rocks and flora) in their habitat [9].

Habitat suitability models of mountain ungulates in the mid hills of western Nepal predict that of the total area studied only 41% is suitable for serow [42]. The study further confirms that suitable habitats for serow were fragmented and mostly confined to the southern part of the study area. Joshi *et al.* in Nepal observed that serow prefers *Quercus semecarpifolia* and *Rhododendron* forests [2]. Studies have shown that serow occupancy is positively correlated with distance to roads, distance to villages, and distance to trails, indicating its preference for areas inaccessible to human encroachment [4]. Across their wide range, serows have been recorded in subalpine, montane, temperate, and tropical forests [4, 43]. Based on isotope analysis, it was suggested that during the Pleistocene, serows in Thailand were more generalist, open-landscape species than the closed-canopy forest-associated species they are now recognized for [44].

Recent discoveries of serow in previously uncharted regions, such as the GunungLedang National Park in Johor, Malaysia, suggest that the serow may persist in isolated parks at lower elevations where poaching is better controlled or may even exhibit signs of recolonization in areas where they were previously thought to be extinct [7]. These newly documented occurrences may contribute to a slight expansion of the known extant range of the serow.

### 2.2 Diet composition of the serow

Knowledge of the diet composition of the Himalayan serow is limited. Giri *et al.* suggested that they tend to be intermediate feeders (between browsers and grazers) consuming grasses, herbs, and woody plants in the central Himalayas during summer [45]. Green in western Himalaya confirmed them to be selective browsers feeding mainly on woody plants, such as oak, forbs, and monocots, forming a very small fraction of its diet [36]. According to some previous studies, it consumes a variety of forage, such as oak leaves, bushes, grasses, shoots, montane bamboo, ferns, moss, and lichen, and, therefore, categorized as a generalist herbivore [46, 47]. They display a preference for nutrient-rich and palatable plant species [35]. Quyoom *et al.*, while studying winter food habits of the Himalayan serow in the temperate forests of Bani Wildlife Sanctuary (western Himalayas), found deciduous shrubs and evergreen oaks as major food components and confirmed them to be browsers at least during winter [22].

Recent shreds of evidence using different approaches such as stable carbon isotopes, Dental mesowear texture analysis (DMTA), Hypsodonty index, and mesowear II have suggested browsing food habits of both Pleistocene and extant *Capricornis* spp. [7, 44, 48]. The DMTA analysis showed that the extant *Capricornis sumatraensis* is a leaf-dominated browser. Moreover, dental adaptations of *Capricornis sumatraensis* also suggest its browsing habits. Detailed investigations on the food habits of this elusive goat using microhistological analysis may aid in understanding the complex ecological perspectives of this mountain ungulate.

### 2.3 Behavior and social organization

Serows are nocturnal, shy, solitary, and little understood. The dearth of information about them is a result of their elusiveness, affinity for rocky terrain, and prevalence in sparse populations [12, 36]. They are crepuscular, peaking in activity in the mornings (06:00–08:00 h) and evenings (16:00–22:00 h) [28, 46]. They hiss and snort when startled, and when disturbed, they yell incoherently [38].

They are typically solitary, although occasionally, male-female couple units or family groups of up to three or four individuals are spotted together [46, 49]. The adult serows of both sexes engage in rigorous intrasexual territoriality, defending their respective domains from rivals of the same sex [49, 50]. Adult serows exhibit a high level of site fidelity as well [50]. Despite solitary roaming, an adult male's range may nearly entirely overlap with an adult female's range, or an adult polygynous male's range may cover two adult female ranges [46, 49].

Possibly associated with territoriality, serows have specific defecation sites, commonly referred to as dung piles [10]. Also used for marking is the secretion of the preorbital gland, which smells like acetic acid and has the consistency of gum arabic [51, 52]. Recent studies on their activity pattern, marking behavior and social interactions show that their activity pattern was bimodal with two peaks, one in the mid-afternoon and the second in late night, whereas resting and ruminating were the highest at noon and twilight. Both sexes used different marking sites and marking frequencies. A total of 33 social behavioral patterns were observed with 18 patterns including agonistic behavior, whereas 15 patterns were relevant to courtship behavior [53].

## 3. Reproduction

Males attain sexual maturity at the age of three, while females achieve it at 2.5 years. The breeding season lasts from October to November, and the young are born in May and June. Twins are rare and the litter size is usually one [38, 46]. There are still some questions to be answered regarding the species' reproductive and mating practices. Here, information on the closely related species *C. crispus* is provided (**Table 1**).

Yearlings start to move away from their mothers' territories but remain somewhat dependent on them. Both sexes depart the natal area when they are between 2 and 4 years old to establish their own territories elsewhere [49, 55]. Female offspring typically stay within their mother's ranges as they create their territories. The adult male is often aggressive and chases away male progeny but is tolerant towards female

| **Age at sexual maturity** | **2–3 years [38]** |
|---|---|
| Breeding season | October-November [38] |
| Gestation period | 215 days [49] |
| Birth season | May-June [38] |
| No. of offspring | Single offspring is common; twins are rare [46] |
| Sex ratio at birth | 1.03:1.00 [49] |
| Weaning age | 6 months [46] |
| Reproductive tenure | 2.5 to 19.5 years [49] |
| Maximum longevity | 21–22 years (females); 20–21 years (males) [49] |

**Table 1.**
*Life history traits of the Serow [54].*

| Location | Total no. of Individuals (MFU) | Living Individuals (MFU) | Time in Captivity (years) | Births (MFU) | Deaths (MFU) |
|---|---|---|---|---|---|
| Aizawl | 1.1.0 | 1.0.0 | 2006–2016 (11) | 0.0.0 | 0.1.0 |
| Assam | 16.6.0 | 3.2.0 | 1979–2016 (38) | 14.3.0 | 12.2.0 |
| Dimapur | 2.1.0 | 2.1.0 | 2013–2016 (04) | 0.0.0 | 0.0.0 |
| Shillong | 1.0.0 | 1.0.0 | 2009–2016 (08) | 0.0.0 | 0.0.0 |

**Table 2.**
*Status of Serow in Indian zoos [54].*

offspring, which explains the various interactions of the adult pair towards the offspring (above 1 year). The mother typically tolerates both genders of offspring, and the close mother-child relationship lasts until the birth of subsequent offspring [49].

### 3.1 Predation

Not much information is available about natural predation on the serow. Some studies have reported the remains of serow in the scats of leopard and red foxes in the Himalayas [14, 35, 56, 57]. Owing to its size, it is unlikely to have many predators other than tiger (*Panthera tigris*) and leopard (*Panthera pardus*). Though the black bear is found in the habitat of Japanese Serow, there is no report of predation by the black bear [58]. The serow responds to disturbance with a low whistling alarm call.

### 3.2 Status in captivity

According to the records that are currently available, 28 captive animals were housed in Indian zoos between 1979 and 2016, with a median of three animals per year. The Central Zoo Authority of India has chosen this species as one of the endangered animal species for the Conservation Breeding Programme. Assam State Zoo and Botanical Garden serves as the Coordinating Zoo for the Conservation Breeding of the Species, with Manipur Zoological Garden serving as a participating institution. The historical population figures of serow kept at various Indian zoos are presented in (**Table 2**).

## 4. Conservation issues

Habitat destruction and resource competition with livestock pose significant threats to serow populations in the Himalayan range, including within protected areas. Habitat fragmentation, alterations in land use, conflicts with predators and local residents, grazing of livestock within serow habitats, poaching, climate change and developmental activities are the major threats to serow [2, 35, 45]. The transmission of diseases such as sarcoptic mange and goat pox is becoming an increasing concern in areas where serow and domestic livestock, especially sheep, coexist.

In the regions of Jammu and Kashmir, Himachal Pradesh, and Uttarakhand, shepherds have come across deceased serow displaying symptoms such as extremely parched skin, internal ruptures and bleeding, loss of fur, and foaming around the mouth [59]. Poaching and infrastructure development, such as the construction of dams and highways, are increasingly responsible for the extensive destruction of habitats, which may serve as important refuges for various species. The presence of livestock in areas where villagers have legal rights to graze their animals and collect timber, and firewood can pose significant challenges in managing competition and potentially modifying the habitat use of species within their range [22].

The detrimental impact of livestock grazing on wild ungulates in the western Himalayan region has been acknowledged as a significant concern within protected areas [60]. Numerous factors contribute to the transmission of diseases, with one notable factor being the seasonal overlap in distribution between wild and domestic ungulates. Notwithstanding, disease transmission can occur in either direction between livestock and wild ungulates through contaminated soil, vegetation and water sources [61]. The potential consequences of disease transmission from livestock to wildlife are of particular concern due to the potential threat it poses to the survival of endangered animal species.

### 4.1 Conclusion and way forward

The Himalayan serow (*Capricornis sumatraensis thar*) is a remarkable species that represents the resilient nature of the Himalayas, demonstrating life's remarkable ability to thrive in some of Earth's most challenging and awe-inspiring environments. The chapter aims to understand this enigmatic species, emphasizing its unique adaptations, its crucial role within the Himalayan ecosystem, and the significant threats it encounters.

The evolutionary brilliance has shaped the Himalayan serow into an animal exquisitely adapted to a mountainous environment. Its robust build, thick fur, and elusive nature are the product of millennia of natural selection, designed to withstand the harsh conditions of the environment. The serow symbolizes resilience, showcasing the incredible capacity of life to adapt to the mountain environment. However, its narrative also serves as an alarming reminder of the environmental challenges we confront, with the serow's very survival being deeply intertwined with habitat conservation, ecosystem health, and the balance of species within the mountains. The challenges faced by the serow also shed light on the broader issues of biodiversity conservation in an era marked by habitat destruction, climate change, and increasing human-wildlife conflicts.

Significant progress has been made in conservation efforts to protect the serow, including the establishment of protected areas, anti-poaching measures, and community involvement, all of which have played crucial roles. Nonetheless, there is still plenty of work to do. Persistent threats such as habitat fragmentation, poaching, and climate change continue to threaten the serow's habitat. Long-term research is needed to better understand the serow's ecology and behavior. There is limited information on the genetic diversity and population structure of Himalayan serow across different geographical regions. This gap hinders effective conservation strategies, especially in the face of habitat fragmentation and changing climates.

Research on the prevalence of diseases and parasites affecting Himalayan serow populations is scarce. Systematic health surveillance could identify emerging threats and help formulate disease management policies. Detailed studies on the diet and nutritional ecology of the Himalayan serow are lacking. Such research would aid in understanding their ecological role and the effects of habitat changes on their feeding behavior and nutritional status. There is a dearth of information on the reproductive strategies, gestation periods, offspring development, and life history traits of the Himalayan serow. Insights into these areas are essential for effective population management and conservation planning. Little is known about the social structure, territoriality, and intraspecific interactions of the Himalayan serow.

Conservation strategies need to be dynamic and adaptable and should be based on collaboration and shared responsibility. Engaging local communities is not only beneficial but essential for long-term conservation success. The plight of the Himalayan serow is an alarming reminder that the conservation of a single species is inherently linked to the preservation of its entire ecosystem, the complex interdependencies within it, and the well-being of the human communities in close proximity. From this review, it becomes clear that the Himalayan serow is not merely a species in peril but a representative of all species facing threats in our rapidly changing world. The story of the serow is a call to action, urging us to ensure that the Himalayas continue to be a refuge for serows and a symbol of life's resilience in challenging conditions. In essence, the future of the serow reflects our dedication to preserving the natural beauty of our planet for future generations.

## Author details

Iyaz Quyoom[1*], Bilal A. Bhat[1] and Wasim Sajad Malik[2]

1 Department of Zoology, University of Kashmir, Hazratbal Srinagar, India

2 Department of Botany, University of Kashmir, Hazratbal Srinagar, India

*Address all correspondence to: iyazquyoom@gmail.com

## References

[1] Chen X, Guan T, Jiang W, Li D, Yang K, Li S. Distribution and population status of bovine species in China based on bibliometric analysis. Biodiversity Science. 2021;**29**(5):668

[2] Joshi R, Basnet DB, Poudel B. Geospatial analysis of habitat suitability for Himalayan Serow *Capricornis sumatraensis* (Bechstein, 1799) in Annapurna conservation area of Nepal by using MaXent model. Iranian Journal of Animal Biosystematics. 2022;**18**(2):121-138

[3] Shepherd CR, Gomez L, Rachmayuningtyas BA. The illegal trade of the Sumatran serow *Capricornis sumatraensis* for traditional medicine in Indonesia. Indonesian Journal of Applied Environmental Studies. 2022;**3**:98-104

[4] Phan TD, Nijhawan S, Li S, Xiao L. *Capricornis sumatraensis*. In: The IUCN Red List of Threatened Species. International Union for Conservation of Nature and Natural Resources; 2020. p. e.T162916735A162916910

[5] Shepherd CR. *Capricornis rubidus*. In: The IUCN Red List of Threatened Species. International Union for Conservation of Nature and Natural Resources; 2021. p. e.T3815A22151023

[6] Mori E, Nerva L, Lovari S. Reclassification of the serows and gorals: The end of a neverending story? Mammal Review. 2019;**49**(3):256-262

[7] Faiznur AA, Yasuda M, Maryati M, Adlil IS, Halid MS. The first record of Sumatran serow, Capricornis sumatraensis (Bovidae, Cetartiodactyla), in Gunung Ledang Johor National Park, a tropical forest remnant on the southern Malay Peninsula. Mammal Study. 2020;**45**(3):259-264

[8] Yadav SK, Lamichhane BR, Thapa SK, Kadariya R, Paudel U, Subedi N, et al. Photographic evidence of Himalayan Serow *Capricornis sumatraensis* ssp. *thar* (Hodgson, 1831) in Bardia National Park, Nepal. Bulletin of Basrah Natural History Museum. 2022;**8**(1):1-9

[9] Aryal A. Status and conservation of Himalayan Serow (*Capricornis sumatraensis*. Thar) in Annapurna conservation area of Nepal. In: BRTF Nepal: A Report Submitted to the Rufford Small Grant for Nature Conservation, UK and the People's Trust for Endangered Species. UK: The Rufford Small Grant Foundation; 2008

[10] Mishra C, Raman T, Johnsingh A. Survey of Primates, Serow, and Goral in Mizoram. Dehra Dun, India: Wildlife Institute of India; 1994

[11] Mishra HR, Mierow D. Wild Animals of Nepal. Ratna Pustak Bhandar; 1976

[12] Schaller GB. Mountain monarchs. In: Wild Sheep and Goats of the Himalaya. Chicago, USA: University of Chicago Press; 1977

[13] Jnawali SR, Baral H, Lee S, Acharya K, Upadhyay G, Pandey M, et al. The status of Nepal mammals: The national red list series, department of national parks and wildlife conservation Kathmandu, Nepal. In: Simon M, editor. Stuart Chair IUCN Species Survival Commission the Status of Nepal's Mammals: The National Red List Series. 4. International Union for Conservation of Nature and Natural Resources; 2011

[14] Aryal A, Sathyakumar S, Kreigenhofer B. Opportunistic animal's diet depend on prey availability: Spring dietary composition of the red

fox (*Vulpes vulpes*) in the Dhorpatan hunting reserve, Nepal. Journal of Ecology and the Natural Environment. 2010;**2**(4):59-63

[15] Wilson DE, Reeder DM. Mammal Species of the World: A Taxonomic and Geographic Reference. Baltimore, Maryland: JHU Press; 2005

[16] Wilson DE, Wilson DE, Mittermeier RA. Handbook of the Mammals of the World, Volume 2: Hoofed Mammals. Barcelona, Spain: Lynx Ediciones; 2011

[17] Liu W, Yao YF, Yu Q, Ni QY, Zhang MW, Yang JD, et al. Genetic variation and phylogenetic relationship between three serow species of the genus Capricornis based on the complete mitochondrial DNA control region sequences. Molecular Biology Reports. 2013;**40**:6793-6802

[18] Dou H, Zhang Y, Feng L. Complete mitochondrial genome of the Himalayan serow (*Capricornis thar*) and its phylogenetic status within the genus Capricornis. Biochemical Systematics and Ecology. 2016;**65**:115-123

[19] Grubb P. Order artiodactyla. Mammal species of the world. In: A Taxonomic and Geographic Reference. 3rd ed. Vol. 1. Johns Hopkins University Press; 2005. pp. 637-722

[20] Sathyakumar S, Sharma LK, Charoo SA. Ecology of Asiatic black bear in Dachigam National Park, Kashmir, India. In: Final Project Report. Dehradun: Wildlife Institute of India; 2013

[21] Bhatnagar YV, Seth CM, Sharma SC, Sharma KK, Gupta M, Singh P. Preliminary assessment of the Himalayan Tahr (*Hemitragus jemlahicus*) in Sarthal-Bani area of Jammu & Kashmir. In: Interim Report by State Forest Research Institute, J&K and Nature Conservation Foundation-ISLT. Mysore; 2007

[22] Quyoom I, Bhat BA, Najar ZH, Tanveer S. Winter diet composition of Himalayan serow *Capricornis sumatraensis thar* in Bani wildlife sanctuary: Implications for the conservation of *Quercus semecarpifolia* forest. Biologia. 2023;**79**(3):853-861

[23] Pandey S. Status and distribution of some Caprids in Himachal Pradesh. In: ENVIS Bulletin: Wildlife and Protected Areas. EurekaMag Biomedical Library; 2002. pp. 30-33

[24] Khanyari M, Geladi I, Ryan R. Observation of Multiple Sarcoptic Mange Related Deaths in Himalayan Serow. Kedarnath Wildlife Sanctuary, India. Gland, Switzerland: IUCN Caprinae Newsletter; 2017

[25] Targe K, Negi HS, Lyngdoh S, Adhikari BS. First sighting of a Himalayan Serow *Capricornis sumatraensis Thar* Hodgson, 1831 in the Cold Desert of Spiti Valley, Himachal Pradesh, India. Journal of the Bombay Natural History Society. 2023;**120**(2)

[26] Pawar D, Singh V, Umariya S, Bakshi S, Antil J, Salaria S, et al. Himalayan serow: Photographic evidence of *Capricornis thar* in Lansdowne Forest division, Uttarakhand, India. Mammal Tales. Zoo's Print. 2018;**33**(12):32-36

[27] Gopal R, Kapoor M et al. Status of Tiger Habitats in High Altitude Ecosystems of Bhutan, India and Nepal (Situation Analysis). New Delhi, Delhi Global Tiger Forum; 2019

[28] Bhattacharya T, Bashir T, Poudyal K, Sathyakumar S, Saha GK. Distribution,

occupancy and activity patterns of goral (*Nemorhaedus goral*) and serow (*Capricornis thar*) in Khangchendzonga biosphere reserve, Sikkim, India. Mammal Study. 2012;**37**(3):173-181

[29] Srivastava T, Kumar A. Seasonal habitat use in three species of wild ungulates in Sikkim Himalaya. Mammalian Biology. 2018;**88**:100-106

[30] Choudhury A. Status of serow (*Capricornis sumatraensis*) in Assam. Tiger Paper. 2003;**30**(2):1-2

[31] Islam N. Record of Himalayan Serow in Manas TR. Indian Forester. 2021;**147**:1030-1033

[32] Wen L, Shi K, Huang J, Song Y, Guo Y. Preliminary analysis of mammal and bird diversity monitored with camera traps in Medog, Tibet. Biodiversity Science. 2014;**22**(6):798

[33] Alliance CC. A preliminary wildlife survey in Sangu-Matamuhuri reserve forest, Chittagong Hill tracts, Bangladesh. In: Unpublished Report Submitted to Bangladesh Forest Department, Dhaka, Bangladesh. 2016

[34] Wang SW. Estimating population densities and biomass of ungulates in the temperate ecosystem of Bhutan. Oryx. 2010;**44**(3):376-382

[35] Aryal A. Habitat ecology of Himalayan serow (*Capricornis sumatraensis* ssp. *thar*) in Annapurna conservation area of Nepal. Tiger Paper. 2009;**34**(4):12-20

[36] Green MJ. Ecological separation in Himalayan ungulates. Journal of Zoology. 1987;**1**(4):693-719

[37] Nijhawan S. Human-animal relations and the role of cultural norms in tiger conservation in the Idu Mishmi of Arunachal Pradesh, India [doctoral dissertation]. London, England: University College London; 2018

[38] Prater SH. The Book of Indian Animals Bombay Natural History Society. Oxford University Press; 1993

[39] Jass CN, Mead JI. *Capricorniscrispus*. Mammalian Species. 2004;**750**:1-10

[40] Thuc PD, BaxterG SC, Hieu DN. Population status of the Southwest China serow *Capricornis milneedwardsii*: A case study in cat Ba Archipelago, Vietnam. Pacific Conservation Biology. 2014;**20**(4):385-391

[41] Buzzard PJ, Li X, Bleisch WV. High altitude ungulate communities in Southwest China. Mammalia. 2018;**82**(5):415-422

[42] Paudel PK, Hais M, Kindlmann P. Habitat suitability models of mountain ungulates: Identifying potential areas for conservation. Zoological studies. 2015;**54**:1-6

[43] Groves C. Systematics of the artiodactyla of China in the 21st century. Zoological Research. 2016;**37**(3):119

[44] Suraprasit K, Jaeger JJ, Shoocongdej R, Chaimanee Y, Wattanapituksakul A, Bocherens H. Long-term isotope evidence on the diet and habitat breadth of Pleistocene to Holocene caprines in Thailand: Implications for the extirpation and conservation of Himalayan gorals. Frontiers in Ecology and Evolution. 2020;**8**:67

[45] Giri S, Aryal A, Koirala RK, Adhikari B, Raubenheimer D. Feeding ecology and distribution of Himalayan serow (*Capricornis thar*) in Annapurna conservation area, Nepal. World Journal of Zoology. 2011;**6**(1):80-85

[46] Sathyakumar S. The Elusive Serow: Surviving under Threat from Humans. Chennai: Frontline; 1997. pp. 71-72

[47] Nowak RM, Paradiso JL. Walker's Mammals of the World. 4th ed. Baltimore: John Hopkins University Press; 1983

[48] Isarankura NA, Ayudhya J, Merceron G, Wannaprasert T, Jaeger JJ, Chaimanee Y, et al. Dental mesowear and microwear for the dietary reconstruction of quaternary southeast Asian serows and gorals. Frontiers in Ecology and Evolution. 2022;**13**(10):1000168

[49] Kishimoto R. Family break-up in Japanese serow, *Capricornis crispus*. In: The Biology and Management of Capricornis and Related Mountain Antelopes. Dordrecht: Springer; 1987. pp. 104-109

[50] Ochiai K, Susaki K. Effects of territoriality on population density in the Japanese serow (*Capricornis crispus*). Journal of Mammalogy. 2002;**83**(4):964-972

[51] Prater SH, Barruel P. The Book of Indian Mammals. Mumbai, India: Bombay Natural History Society; 1971

[52] Soma H. Serows (genus *Capricornis*). In: Grzimek's Encyclopedia of Mammals. Vol. 5. New York: McGraw Hill Publishing Company; 1990. pp. 505-509

[53] Lovari S, Mori E, Procaccio EL. On the behavioural biology of the mainland serow: A comparative study. Animals. 2020;**10**(9):1669

[54] National Studbook of Himalayan Serow (*Capricornis thar*), Wildlife Institute of India, Dehradun and Central Zoo Authority, New Delhi. TR. No.2016/008. p. 4

[55] Ochiai K. Pair-bond and mother-offspring relationships of Japanese serow in Kusoudomari, Wakinosawa Village. Journal of the Mammalogical Society of Japan. 1983;**9**:192-203

[56] Shah GM, Jan U, Bhat BA, Ahmad F, Ahmad J. Food habits of the leopard *Panthera pardus* in Dachigam National Park, Kashmir, India. Journal of Threatened Taxa. 2009;**26**:184-185

[57] Habib B, Gopi GV, Noor A, Mir ZR. Ecology of Leopard *Panthera pardus* in relation to prey abundance and land use pattern in Kashmir Valley. In: Project Completion Report Submitted to Department of Science and Technology, Govt. of India. Vol. 6. Dehra Dun, India: Wildlife Institute of India; 2014

[58] Ochiai K, Nakama S, Hanawa SI, Amagasa T. Population dynamics of Japanese serow in relation to social organization and habitat conditions. I. Stability of Japanese serow density in stable habitat conditions. Ecological Research. 1993;**8**(1):11-18

[59] Khanyari M, Geladi I, Rachael R. Observation of Multiple Sarcoptic Mange Related Deaths in Himalayan Serow, in Kedarnath Wildlife Sanctuary, India. Gland, Switzerland: Newsletter for the Caprinae Specialist Group; 2012

[60] Bhat RA, Tak H, Bhat BA, Dar JA, Ahmad R. Gastrointestinal helminth parasites of wild ungulates in Hirpora wildlife sanctuary, Kashmir, India. Journal of Parasitic Diseases. 2022;**46**(3):804-810

[61] Nunn CL, Thrall PH, Leendertz FH, Boesch C. The spread of fecally transmitted parasites in socially-structured populations. PLoS One. 2011;**6**(6):e21677

Chapter 7

# Dark Diversity Bringing Light to Forest Science

*Magno Gonçalves-Araújo*


## Abstract

Dark diversity offers a transformative perspective in forest science by addressing species that are absent but ecologically viable within ecosystems. Complementing observed diversity, it forms the species pool, providing a more holistic understanding of biodiversity. This concept is essential for conservation, particularly in fragmented and degraded landscapes, where it identifies ecological barriers to species reestablishment, such as dispersal limitations or habitat degradation. By incorporating dark diversity, restoration projects can prioritize areas with high biodiversity potential, optimize species reintroductions, and enhance ecological resilience. Furthermore, dark diversity supports in designing ecological corridors, improving functional stability, and addressing challenges like climate change and habitat fragmentation. While its practical application is emerging, advances in ecological modeling and local knowledge integration are making dark diversity a valuable tool for guiding biodiversity conservation and forest ecosystem restoration. This framework bridges theory and practice, offering a critical methodology for sustaining global biodiversity and ecosystem functionality.


**Keywords:** dark diversity, biodiversity, forest restoration, ecosystem resilience, species pool

## 1. Introduction

Understanding biodiversity requires a comprehensive approach that transcends traditional metrics. Observed diversity, which quantifies the species present in a given area, has long been the most commonly used metric to quantify biodiversity [1, 2]. However, this metric alone provides an incomplete picture of biodiversity, as it excludes species that could potentially inhabit the area under suitable environmental and ecological conditions. This "missing" set of species is known as dark diversity [3]. The integration of observed diversity and dark diversity forms the basis of the species pool, representing the total assemblage of species that are either present or ecologically viable in a given habitat [4–6], where ecologists can achieve a more nuanced understanding of biodiversity. This holistic perspective is essential for developing effective strategies to conserve and restore ecosystems in a rapidly changing world.

In forest ecosystems worldwide, dark diversity provides critical insights into the underlying drivers of species absence, such as dispersal limitations, habitat degradation, or biotic interactions. Its application in forest ecology allows researchers to assess the "hidden" potential of biodiversity, particularly in areas undergoing significant anthropogenic pressures like deforestation, fragmentation, and climate change [1]. For instance, in tropical and subtropical forests, where biodiversity is exceptionally high but under severe threat, integrating dark diversity into conservation planning helps identify species at risk of local extinction and prioritize restoration actions to enhance ecological resilience [1, 7, 8]. By highlighting the unrealized biodiversity potential, dark diversity offers a novel framework to refine forest management strategies, optimize species reintroductions, and foster ecosystem stability in forested landscapes worldwide.

Forests represent some of the most biodiverse ecosystems on Earth, playing a vital role in global ecological stability and providing a multitude of ecosystem services [9, 10]. However, ongoing anthropogenic pressures such as habitat loss, fragmentation, and climate change are causing alarming declines in biodiversity [1, 11–16]. Traditional biodiversity metrics, such as observed diversity, have been invaluable for understanding ecological patterns and processes. Yet, they often overlook the potential pool of species that could thrive in a given area under suitable conditions. This concept, termed "dark diversity," offers a novel perspective by encompassing the species absent from an area despite favorable environmental conditions, thereby providing insights into ecological community assembly and resilience [1, 3, 17].

In forest science, dark diversity holds the potential to revolutionize our understanding of biodiversity dynamics [3]. By bridging the gap between observed diversity and the potential species pool, it can reveal the hidden ecological processes that shape community composition and functionality. This approach has significant implications for forest conservation and management, especially in light of increasing global environmental changes. For instance, integrating dark diversity into studies of forest fragmentation can shed light on the impacts of habitat isolation and degradation on species pools, thereby guiding more effective conservation strategies [1, 7, 18–20].

The study of dark diversity is critical for understanding biodiversity dynamics in tropical forests, particularly in highly diverse and threatened regions like Brazil, offering insights into the potential biodiversity that could be supported under optimal ecological conditions [1]. Furthermore, understanding dark diversity can enhance forest restoration practices by identifying species that are absent yet ecologically compatible with a given site. Despite its promise, the concept of dark diversity remains underexplored in forest science, particularly in tropical and subtropical ecosystems, where biodiversity levels are highest and threats are most pronounced.

In tropical forests, where deforestation, habitat fragmentation, and climate change drastically alter species distributions, dark diversity provides a framework to assess ecological debt and restoration potential. For Brazil, a global biodiversity hotspot, this perspective is invaluable for informing conservation and restoration strategies. By integrating dark diversity into ecological assessments, it becomes possible to identify missing species, enabling the development of interventions that prioritize not only existing biodiversity but also the ecological potential of restored landscapes [1, 7, 18]. Such approaches are crucial for mitigating the loss of ecosystem services and ensuring the long-term resilience of tropical forests in the Anthropocene.

This chapter aims to explore the role of dark diversity in advancing forest science, focusing on its applications in biodiversity assessment, conservation planning, and ecosystem restoration. By integrating dark diversity into forest research, we can illuminate the unseen dimensions of biodiversity, thereby fostering a deeper understanding of forest ecosystems and enabling more informed ecological management practices.

### 1.1 Role of dark diversity in biodiversity assessment

Integrating dark diversity into biodiversity assessments enables the identification of gaps between potential and observed diversity, enhancing our understanding of the limiting factors for species presence. Studies have demonstrated that dark diversity is influenced by abiotic conditions, such as resource availability and soil quality, as well as biotic factors, including competition and mutualistic interactions [18, 21]. For instance, in European forest ecosystems, the combination of observed and dark diversity data has underscored the importance of connectivity among forest fragments for biodiversity conservation, highlighting how habitat fragmentation limits the realization of species pools [17].

The application of dark diversity is equally promising in tropical and subtropical ecosystems, which harbor the planet's highest biodiversity levels but also face the most significant threats. The Caatinga dry forest, for example, is characterized by a rich diversity of tree and shrub species, many of which currently exist in dark diversity due to intense logging and deforestation [1]. In this context, dark diversity serves as a valuable tool for identifying priority species for conservation and guiding ecological restoration initiatives. Therefore, using dark diversity in biodiversity assessments is critical for predicting how communities will respond to environmental changes. This approach not only provides insights into the ecological processes shaping community assembly but also supports ecosystem management strategies, promoting resilience and ecological functionality [7, 18, 22–25].

In forest science, incorporating dark diversity has led to the revision of monitoring and management strategies. For example, in the forests of China, a study has identified key areas where conservation efforts should be concentrated, given their potential to support both currently present and potentially recolonizing species, which are crucial for maintaining ecological balance and ecosystem services [26]. Dark diversity is also instrumental in shaping environmental policies [27]. By capturing the unrealized biotic potential of degraded areas, it helps identify targets for restoration and conservation prioritization, while providing critical information for designing ecological corridors and protected area networks. Thus, dark diversity is more than an additional metric; it is a transformative approach that sheds light on ecosystems' hidden potential and contributes to building a more sustainable future for global biodiversity.

The incorporation of dark diversity into biodiversity assessments represents a significant advancement in our understanding of ecological dynamics and species distribution [28]. In the context of global biodiversity evaluation, dark diversity plays a crucial role in elucidating the factors that limit species establishment and persistence. By accounting for both observed and dark diversity, researchers can better understand the processes driving community assembly, including the effects of competition, dispersal limitation, and environmental change [1, 29]. This dual approach provides a more holistic view of biodiversity, particularly in regions facing significant anthropogenic pressures and environmental change.

Furthermore, dark diversity has the potential to inform conservation strategies by identifying areas with high species potential, even if those species are currently absent. This predictive capability is essential for developing proactive conservation measures, ensuring that ecosystems can maintain or recover their biodiversity in the face of global challenges such as climate change and habitat degradation. As such, dark diversity is an indispensable tool for enhancing the accuracy and effectiveness of biodiversity assessments and for informing policies aimed at preserving global biodiversity. Integrating dark diversity into global biodiversity assessments not only enriches our understanding of species potential but also strengthens our ability to predict and mitigate biodiversity loss. It is a powerful metric for enhancing ecological management and conservation efforts, providing critical insights into the future of biodiversity in a rapidly changing world.

This approach is particularly relevant in tropical forests, such as those in Brazil, where habitat fragmentation, deforestation, and climate change decrease species richness. By coupling dark diversity with traditional biodiversity metrics, researchers can uncover the full potential of species pools, enabling more accurate predictions of ecosystem resilience [1]. Such insights are vital for prioritizing conservation actions in tropical regions, ensuring the diversity richness and the services these ecosystems provide to both local and global communities.

### 1.2 Role of dark diversity in ecosystem conservation planning

Ecosystem conservation planning is one of the central pillars of biodiversity preservation, aiming to maintain ecological functions and prevent further species loss. Dark diversity is crucial for understanding the full potential of an ecosystem, as it accounts for species that could recolonize an area following habitat restoration or reduced anthropogenic pressure. This approach is especially relevant in fragmented or degraded landscapes, where observed species diversity may not fully represent the ecological potential of a region. As climate change, habitat fragmentation, and land-use change continue to alter ecosystems globally, dark diversity provides essential insights into the dynamic nature of species distributions and the capacity of ecosystems to recover and adapt [1, 26, 28]. One of the fundamental contributions of dark diversity to conservation planning is its ability to highlight areas with untapped biodiversity potential. By identifying regions with a high potential for species establishment, conservationists can prioritize restoration efforts in areas that may have previously been overlooked. For example, studies have shown that landscapes with high levels of dark diversity, even when current species richness is low, can harbor significant restoration potential [1, 7, 18, 26]. These areas are often key in ensuring the resilience of ecosystems, as they provide the opportunity for species to reestablish themselves as environmental conditions change.

Moreover, dark diversity plays a vital role in predicting the success of species restoration. Species that are absent but ecologically suited to a given area may be unable to establish due to dispersal limitations or biotic interactions, such as competition. Understanding the underlying mechanisms that prevent species from filling available ecological niches is critical for developing effective conservation strategies. For example, recent studies have demonstrated that dark diversity can be influenced by both biotic and abiotic factors, such as habitat connectivity and climate [1, 29]. By integrating dark diversity metrics into conservation planning, practitioners can identify the factors that may limit species reestablishment and implement targeted interventions to overcome these barriers.

In the context of landscape ecology, dark diversity provides a more nuanced understanding of biodiversity patterns across spatial scales. Traditional conservation planning often focuses on individual species or specific habitat types, but this approach may fail to capture the broader ecological context in which species interact. Dark diversity, by considering species that are not currently present, enables a more holistic view of biodiversity, emphasizing the role of landscape connectivity and habitat heterogeneity in supporting ecosystem functions [1, 29]. For instance, areas with high dark diversity are often more resilient to environmental changes because they are more likely to harbor species with a range of ecological tolerances and functional traits [30, 31]. This perspective is particularly important when addressing global challenges such as climate change, where shifts in species distributions are expected to occur rapidly and unpredictably.

Incorporating dark diversity into ecosystem conservation planning also has significant implications for policy development. As governments and conservation organizations strive to meet global biodiversity targets, such as those outlined in the Convention on Biological Diversity (CBD; https://www.cbd.int) and the Sustainable Development Goals (SDGs; https://sdgs.un.org/goals), it is essential to consider both the current and potential biodiversity of ecosystems [27]. Dark diversity metrics provide a more comprehensive approach to biodiversity monitoring, allowing policymakers to better assess the long-term viability of conservation strategies and to prioritize areas for protection and restoration that offer the greatest potential for species recovery [26, 27, 32]. For example, in forest ecosystems, research has shown that areas with high dark diversity, even in the face of significant degradation, may be restored more effectively through targeted management interventions [26].

Furthermore, the integration of dark diversity into conservation planning aligns with a growing recognition of the need for adaptive management practices. As ecosystems continue to be impacted by human activities and environmental change, conservation strategies must remain flexible and responsive to shifting ecological conditions. Dark diversity offers a framework for anticipating and mitigating the effects of these changes, providing early warnings of potential shifts in species distributions and ecosystem functions [1, 7, 18]. By acknowledging the role of dark diversity in shaping biodiversity trajectories, conservationists can develop more resilient and sustainable management strategies that account for both current and future ecological needs.

In tropical regions, for example, Caatinga dry forest, where deforestation and habitat fragmentation have disrupted ecological processes, dark diversity can guide conservation efforts by prioritizing areas with high potential for biodiversity recovery [1]. By integrating dark diversity into planning, conservation strategies become more holistic, focusing not only on protecting existing species but also on restoring the ecological of degraded landscapes, thus ensuring the long-term persistence of tropical biodiversity.

### 1.3 Role of dark diversity in ecosystem restoration

Ecosystem restoration aims to recover biodiversity and ecosystem services in degraded landscapes by addressing both present and potential ecological processes [33–35]. Incorporating dark diversity into restoration planning offers an opportunity to expand the scope of conservation efforts by identifying species that could enhance the ecological resilience and functionality of restored systems. This approach shifts the focus from merely replacing lost species to fostering ecological networks that support long-term sustainability [7, 18, 31]. Dark diversity provides insights into the unrealized potential of ecosystems, particularly in highly fragmented landscapes. For instance, fragmented habitats often suffer from an extinction debt—a delayed loss of species due to historical fragmentation—which can be reversed by reintroducing species from the local species pool and that was in the dark diversity [1, 22].

A significant advantage of integrating dark diversity into restoration efforts is its diagnostic potential. Understanding why certain species are missing from a habitat can illuminate ecological barriers to species establishment. Factors such as dispersal limitations, soil degradation, or competition with invasive species often hinder the natural recolonization of species that are in the dark diversity [1, 7, 28]. For example, in restoration projects, reintroduction of species from that were in the dark diversity has been shown to increase plant species richness and functional diversity, even in highly degraded systems [7, 31]. Identifying and mitigating these barriers through targeted interventions—such as soil amendments, removal of invasive species, or facilitation of seed dispersal—can significantly enhance the success of restoration projects.

Restoration projects using dark diversity have a particular relevance for forest restoration, as it enables practitioners to move beyond the visible loss of biodiversity and address the underlying ecological processes that limit species reestablishment. By incorporating dark diversity, restoration efforts can be more effectively tailored to promote functional resilience and ecological coherence, addressing challenges such as species dispersal barriers and competition with invasive flora [1, 7, 28, 31]. Furthermore, understanding dark diversity enriches ecological theory by linking community assembly processes to conservation practice, emphasizing the role of latent biodiversity in sustaining ecosystem services.

Despite its promise, the practical application of dark diversity in ecosystem restoration remains still very little studied. Methodological challenges, such as accurately identifying species within the dark diversity and understanding their ecological requirements, must be addressed. Advances in ecological modeling and metacommunity theory have facilitated progress in this area by providing tools to predict the compatibility of dark diversity species with specific habitats [3, 32]. Furthermore, integrating local knowledge and historical records can improve the accuracy of dark diversity assessments, particularly in regions where baseline biodiversity data are limited. As restoration ecology continues to evolve, embracing the concept of dark diversity will enable more effective strategies to combat biodiversity loss and promote ecosystem resilience in the face of global change.

In tropical forest regions, integrating the concept of dark diversity into restoration strategies is essential for advancing ecological restoration and biodiversity conservation [1, 3, 7]. By identifying missing species and their functional roles, restoration plans can prioritize interventions that enhance ecological resilience and reestablish disrupted ecosystem functions [18, 30, 31]. In tropical ecosystems, where deforestation and fragmentation have drastically altered biodiversity and ecosystem dynamics, addressing dark diversity allows for more targeted restoration efforts that aim not only to recover lost species but also to rebuild the functional and structural integrity of forested landscapes [1]. Incorporating this concept into restoration practices not only improves the ecological integrity of restored landscapes but also ensures their long-term resilience of ecosystems. By aligning restoration goals with the ecological potential revealed by dark diversity, it becomes possible to create biodiverse, resilient ecosystems capable of withstanding ongoing environmental changes.

*DOI: http://dx.doi.org/10.5772/intechopen.1008966*

## 2. Conclusion

The concept of dark diversity represents a transformative advancement in landscape ecology and forest science, offering a novel methodology for the development of ecological indicators. By identifying species absent but compatible with specific habitats, dark diversity enables a more nuanced understanding of ecosystem potential and constraints. Its application is vital for enhancing biodiversity, guiding conservation strategies, and informing restoration projects by bridging the gap between theoretical ecology and practical interventions. Incorporating dark diversity into landscape management frameworks not only improves the effectiveness of restoration but also strengthens efforts to conserve ecological integrity and functional resilience in the face of escalating environmental pressures. As such, dark diversity serves as a critical tool for advancing both the science and practice of biodiversity conservation and forest ecosystem restoration.

Understanding and integrating the concept of dark diversity into biodiversity and forest science is pivotal for advancing ecological research and conservation strategies. By shedding light on the species that are ecologically compatible with a habitat yet absent due to anthropogenic or natural barriers, dark diversity provides a deeper comprehension of community assembly processes and ecosystem resilience. In tropical forests and unique ecosystems such as the Caatinga, acknowledging dark diversity can reveal hidden patterns of species loss and highlight restoration opportunities.

From a forest science perspective, incorporating dark diversity metrics enhances our ability to prioritize conservation actions and set more realistic restoration targets. This approach bridges ecological theory with practical applications, enabling scientists and policymakers to design interventions that not only restore biodiversity but also strengthen ecosystem services. As well, dark diversity enriches our understanding of ecological potential, offering a novel dimension for safeguarding the integrity and functionality of forested landscapes.

The integration of dark diversity into ecosystem restoration strategies represents a transformative approach to understanding and addressing biodiversity loss. By identifying species that are ecologically compatible yet absent from degraded habitats, dark diversity provides a roadmap for targeted restoration efforts that go beyond merely increasing species richness. It allows restoration practitioners to prioritize species and functional groups that contribute to the resilience and stability of ecosystems, particularly in biodiversity hotspots such as tropical forests and the Caatinga.

For restoration ecology, incorporating dark diversity into planning ensures a more comprehensive understanding of the ecological potential of degraded landscapes. This perspective not only supports in the selection of species that could thrive under restored conditions but also highlights the underlying barriers to their establishment, such as habitat fragmentation, dispersal limitations, or ecological mismatches. By addressing these barriers, restoration initiatives can more effectively reestablish ecological networks and recover critical ecosystem functions.

Finally, the inclusion of dark diversity shifts the focus from what is present to what could be achieved, aligning restoration goals with the ecological and evolutionary potential of ecosystems. This framework offers a valuable tool for advancing restoration science and fostering more effective, science-driven strategies to counteract biodiversity loss and ecosystem degradation in a rapidly changing world.

## Author details

Magno Gonçalves-Araújo
Federal University of Ceará, Ceará, Brazil

*Address all correspondence to: magno_mm2006@yahoo.com.br

## References

[1] de Oliveira Gonçalves-Araújo MD, de Carvalho CE, Pequeno PACL, et al. Observed and dark diversity of plants' life-forms are driven by climate and human impacts in a tropical dry forest. Biodiversity and Conservation. 2024;**33**:759-773. DOI: 10.1007/s10531-023-02771-z

[2] Wiens JJ. The causes of species richness patterns across space, time, and clades and the role of "ecological limits." The Quarterly Review of Biology. 2011;**86**(2):75-96. DOI: 10.1086/659883

[3] Pärtel M, Szava-Kovats R, Zobel M. Dark diversity: Shedding light on absent species. Trends in Ecology & Evolution. 2011;**26**(3):124-128. DOI: 10.1016/j.tree.2010.12.004

[4] Pärtel M, Zobel M, Zobel K, van der Maarel E, Partel M. The species Pool and its relation to species richness: Evidence from Estonian plant communities. Oikos. 1996;**75**(1):111. DOI: 10.2307/3546327

[5] Zobel M, van der Maarel E, Dupré C. Species pool: The concept, its determination and significance for community restoration. Applied Vegetation Science. 1998;**1**(1):55-66. DOI: 10.2307/1479085

[6] Zobel M. The species pool concept as a framework for studying patterns of plant diversity. Journal of Vegetation Science. 2015;**27**(1):8-18. DOI: 10.1111/jvs.12333

[7] Lewis RJ, de Bello F, Bennett JA, Fibich P, Finerty GE, Götzenberger L, et al. Applying the dark diversity concept to nature conservation. Conservation Biology. 2016;**31**(1):40-47. DOI: 10.1111/cobi.12723

[8] Trindade DPF, Carmona CP, Reitalu T, Pärtel M. Observed and dark diversity dynamics over millennial time scales: Fast life-history traits linked to expansion lags of plants in Northern Europe. Proceedings of the Royal Society of London. Series B, Biological Sciences. 2023;**290**:20221904. DOI: 10.1098/rspb.2022.1904

[9] Brockerhoff EG, Barbaro L, Castagneyrol B, et al. Forest biodiversity, ecosystem functioning and the provision of ecosystem services. Biodiversity and Conservation. 2017;**26**:3005-3035. DOI: 10.1007/s10531-017-1453-2

[10] Chen S, Chen J, Jiang C, Yao RT, Xue J, Bai Y, et al. Trends in research on Forest ecosystem Services in the Most Recent 20 years: A bibliometric analysis. Forests. 2022;**13**:1087. DOI: 10.3390/f13071087

[11] Fahrig L. Ecological responses to habitat fragmentation per Se. Annual Review of Ecology, Evolution, and Systematics. 2017;**48**(1):1-23. DOI: 10.1146/annurev-ecolsys-110316-022612

[12] Fletcher RJ, Didham RK, Banks-Leite C, Barlow J, Ewers RM, Rosindell J, et al. Is habitat fragmentation good for biodiversity? Biological Conservation. 2018;**226**:9-15. DOI: 10.1016/j.biocon.2018.07.022

[13] Hanski I. Habitat fragmentation and species richness. Journal of Biogeography. 2015;**42**(5):989-993. DOI: 10.1111/jbi.12478

[14] Mantyka-pringle CS, Martin TG, Rhodes JR. Interactions between climate and habitat loss effects on biodiversity: A systematic review and meta-analysis. Global Change Biology. 2011;**18**(4):1239-1252. DOI: 10.1111/j.1365-2486.2011.02593.x

[15] Püttker T, Crouzeilles R, Almeida-Gomes M, Schmoeller M, Maurenza D, Alves-Pinto H, et al. Indirect effects of habitat loss via habitat fragmentation: A cross-taxa analysis of forest-dependent species. Biological Conservation. 2020;**241**:108368. DOI: 10.1016/j.biocon.2019.108368

[16] Villard M-A, Metzger JP. REVIEW: Beyond the fragmentation debate: A conceptual model to predict when habitat configuration really matters. Journal of Applied Ecology. 2014;**51**(2):309-318. DOI: 10.1111/1365-2664.12190

[17] Pärtel M, Szava-Kovats R, Zobel M. Community completeness: Linking local and dark diversity within the species pool concept. Folia Geobotanica. 2013;**48**(3):307-317. DOI: 10.1007/s12224-013-9169-x

[18] Moeslund JE, Brunbjerg AK, Clausen KK, Dalby L, Fløjgaard C, Juel A, et al. Using dark diversity and plant characteristics to guide conservation and restoration. Journal of Applied Ecology. 2017;**54**(6):1730-1741. DOI: 10.1111/1365-2664.12867

[19] Trindade DPF, Carmona CP, Pärtel M. Temporal lags in observed and dark diversity in the anthropocene. Global Change Biology. 2020:1-9. DOI: 10.1111/gcb.15093

[20] Trindade DPF, Pärtel M, Carmona CP, Randlane T, Nascimbene J. Integrating dark diversity and functional traits to enhance nature conservation of epiphytic lichens: A case study from Northern Italy. Biodiversity and Conservation. 2021;**30**(8-9):2565-2579. DOI: 10.1007/s10531-021-02211-w

[21] Riibak K, Reitalu T, Tamme R, et al. Dark diversity in dry calcareous grasslands is determined by disper sal ability and stress-tolerance. Ecography. 2015;**38**:713-721. DOI: 10.1111/ecog.01312

[22] Riibak K, Noreika N, Helm A, et al. Plants, fungi, and carabid beetles in temperate forests: Both observed and dark diversity depend on habitat availability in space and time. Landscape Ecology. 2024;**39**:158. DOI: 10.1007/s10980-024-01960-7

[23] Ronk A, Szava-Kovats R, Pärtel M. Applying the dark diversity concept to plants at the European scale. Ecography. 2015;**38**(10):1015-1025. DOI: 10.1111/ecog.01236

[24] Ronk A, de Bello F, Fibich P, Pärtel M. Large-scale dark diversity estimates: New perspectives with combined methods. Ecology and Evolution. 2016;**6**(17):6266-6281. DOI: 10.1002/ece3.2371

[25] Wani SA, Ahmad R, Dar FA, et al. Estimating dark diversity and regional species pool in the high-altitude Himalayan habitats. Biodiversity and Conservation. 2023;**32**:2981-2999. DOI: 10.1007/s10531-023-02639-2

[26] Tang L, Wang R, He KS, Shi C, Yang T, Huang Y, et al. Throwing light on dark diversity of vascular plants in China: Predicting the distribution of dark and threatened species under global climate change. PeerJ. 2019;**7**:e6731. DOI: 10.7717/peerj.6731

[27] Cliquet A, Decleer K. Halting and restoring species loss: Incorporating the concepts of extinction debt, ecological trap and dark diversity into conservation and restoration law. Griffith Law Review. 2017;**26**(2):178-201. DOI: 10.1080/10383441.2017.1355873

[28] Paganeli B, Toussaint A, Bueno CG, Fujinuma J, Reier Ü, Pärtel M. Dark diversity at home describes the success of cross-continent tree invasions. Diversity

and Distributions. 2022;**28**:1202-1213. DOI: 10.1111/ddi.13522

[29] Fløjgaard C, Valdez JW, Dalby L, Moeslund JE, Clausen KK, Ejrnæs R, et al. Dark diversity reveals importance of biotic resources and competition for plant diversity across habitats. Ecology and Evolution. 2020:1-11. DOI: 10.1002/ece3.6351

[30] Deschênes É, Santala KR, Lavigne J, Aubin I. Using a trait-based dark diversity approach to evaluate natural recovery potential in forests. Restoration Ecology. 2024;**32**:e14251. DOI: 10.1111/rec.14251

[31] Morel L, Jung V, Chollet S, Ysnel F, Barbe L. From taxonomic to functional dark diversity: Exploring the causes of potential biodiversity and its implications for conservation. Journal of Applied Ecology. 2022;**59**:103-116. DOI: 10.1111/1365-2664.14033

[32] Fujinuma J, Pärtel M. Decomposing dark diversity affinities of species and sites using Bayesian method: What accounts for absences of species at suitable sites? Methods in Ecology and Evolution. 2023;**14**:1796-1807. DOI: 10.1111/2041-210X.14109

[33] Bullock JM, Aronson J, Newton AC, Pywell RF, Rey-Benayas JM. Restoration of ecosystem services and biodiversity: Conflicts and opportunities. Trends in Ecology & Evolution. 2011;**26**(10):541-549. DOI: 10.1016/j.tree.2011.06.011

[34] Hua F et al. The biodiversity and ecosystem service contributions and trade-offs of forest restoration approaches. Science. 2022;**376**:839-844. DOI: 10.1126/science.abl4649

[35] Rey Benayas JM, Bullock JM. Restoration of biodiversity and ecosystem services on agricultural land. Ecosystems. 2012;**15**(6):883-899. DOI: 10.1007/s10021-012-9552-0

Chapter 8

# What Bioclimatic Factors Drive Habitat Suitability and Spatial Distribution of the Afrotemperate Forest Patches in the Mountainous Grassland Park in South Africa?

*Mahlomola Ernest Daemane, Samuel Adelabu and Abel Ramoelo*

## Abstract

The forests across the globe are undergoing significant changes because of climate change, which is mainly accountable for the frequency and intensity of extreme weather events. Nevertheless, numerous forests like those in the Golden Gate Highlands National Park (GGHNP) are crucial biogeographical connections between extensive forest regions in southern Africa, offering appropriate habitats for numerous forest species. This research employed bioclimatic variables to forecast the historical distribution of Afrotemperate forests in South Africa, encompassing the isolated patches in the GGHNP. The findings from the current study indicated that the optimal habitat for Afrotemperate forest species is found along the Drakensberg Mountain range in South Africa. Annual mean temperature, annual precipitation, the precipitation of the wettest month, the precipitation of the wettest quarter, the precipitation of the warmest quarter, and the precipitation of the coldest quarter were the key climatic factors affecting the spatial distribution of the Afrotemperate forest. The peak frequency for rainfall in the wettest month and quarter ranged from 87.4 to 174.8 mm and 225.6 to 451.2 mm, respectively. The peak occurrence of annual average temperature ranged from 12.9 to 18.2 °C. The area under the curve of the receiver operating characteristics (ROC) for both Bioclim and Random Forest exhibited good predictive performance and fit, with AUC values of 0.982 and 0.977, respectively. Understanding the effects of climate as well as the various levels of response to climatic variation is therefore critical to the conservation of Afrotemperate forest.

**Keywords:** Afrotemperate, species distribution model, biodiversity, habitat suitability, conservation

## 1. Introduction

High-altitude ecosystems are renowned for their remarkable biodiversity, featuring a diverse range of plant species that have adjusted to harsh climatic conditions, which makes them an important focus for research and conservation initiatives [1–4]. Mountains are defined by their significant topographical variation and cooler climates, where the temperature drops approximately 0.5–0.6°C for each 100 m of elevation gained relative to the nearby lowlands [5]. Furthermore, mountains undergo fewer human impacts compared to the lowlands and may serve as sanctuaries for both plants and animals [3, 5, 6]. Nonetheless, Afrotemperate forests are often found in regions with high human population density and face significant threats from agriculture, fire, and grazing [7, 8].

The climate differences observed across the various mountain ranges in southern Africa significantly influence the ecosystem and species living in these areas because of variations in longitude, topography, size, elevation, and aspect [9]. Afrotemperate forests are typically small and fragmented, often found within a grassland matrix, and usually support a variety of endemic species that cannot migrate to other suitable habitats as the climate warms [7, 8]. Forests are found in regions with seasonal temperature changes, primarily in areas with higher rainfall or where soil moisture remains adequate for long durations [10]. As elevations rise, temperatures drop, leading to a noticeable temperature gradient at various altitudes, which affects vegetation patterns, species distribution, and overall biodiversity in mountainous environments [9]. Climate change may also cause alterations and reductions in species distribution [11, 12]. Information regarding the current spatial distribution of forest species, the impact of climate change, and the potential growth of their suitable habitats is essential for identifying threatened ecosystems and possible niches [13].

Changes in environmental drivers and climatic fluctuations result in altered plant composition and structure over time [14]. As a result, many species with similar environmental factors and climates live in similar habitats [14]. Forest ecosystems are constantly changing; sometimes, they undergo recurring catastrophic disturbances that completely reorganize the ecosystem. This can encourage the emergence of a new ecosystem with a different variety of species [15]. Both short-term and long-term vegetation analyses conducted as part of global change studies have revealed significant changes in plant diversity [1, 16]. Moreover, severe weather fluctuations and wildfires in mountainous areas can also impact essential ecosystem services, changing the amount of water available and affecting the make-up and structure of forest and woodland communities found within a grassland matrix [17, 18]. Nevertheless, certain forest ecosystems maintain stability for numerous decades under natural conditions, with changes occurring at a considerably smaller scale [19]. In the Drakensberg Mountains of South Africa, ongoing investigations are examining the spatial distribution of the Afrotemperate forest and the environmental impacts affecting these forest islands within a grassland matrix [15, 17–20].

Species distribution models (SDMs) are being used more and more to identify areas for biodiversity conservation, including protected and threatened species, to determine these changes in habitat suitability [13, 21–23]. Additionally, they are necessary for evaluating the possible impacts of climate change on the habitats and distribution patterns of species [23]. Because of climate change, many forest species that are already flourishing in their habitats may also show some adaptability outside of their natural range [13]. Nevertheless, many studies on forests under climate change ignore this flexibility, estimating species' climate needs based only on their natural ranges and applying climate change scenarios [13]. Reliable predictions of future changes and accurate and precise estimates of current ecological conditions are essential for making well-informed and effective management decisions [24].

The evaluation of predictive accuracy holds equal significance in the implementation of species distribution modeling techniques [21, 22, 25–27]. The juxtaposition of techniques permits researchers to examine how varying attributes of the data and the species influence the accuracy of the predictive maps produced by the model [27, 28]. Our prior investigation on forest and woodland mapping within the GGHNP established that generalized linear model (GLM) analysis serves as an effective instrument in species distribution modeling; however, the impacts of bioclimatic variables were not incorporated [20, 29].

In this study, the Bioclim and Random Forest models were used to predict the spatial distribution of Afrotemperate species using 19 bioclimatic variables extracted from the DIVA-GIS 7.5 software [30]. The Bioclim variables are used in approximately 76% of recently published Maxent analytical ecological studies that assess the potential effects of climate change and conservation biogeography concerns [13, 23, 30–32]. The Random Forest algorithm has also gained popularity in remote sensing [29, 33], forestry [34], ecology [35–37], and climate change [38, 39]. Random Forest is robust to noise, even in the presence of many independent variables [40, 41]. As a result, these models have demonstrated significant efficacy as tools for predicting species spatial distribution. The results of this study will hence contribute to the conservation of Afrotemperate forests and furnish essential habitat and species occurrence data amidst the challenges posed by climate change.

## 2. Materials and methods

### 2.1 Study area

With a focus on the GGHNP, this study examined the distribution of Afrotemperate forest species in South Africa. Only a small portion of the park's overall area is made up of isolated forests. The GGHNP is situated in South Africa's eastern Free State Province, at the base of the Drakensberg Mountains. The GGHNP is located on the border between South Africa and Lesotho, between latitudes 28,030′ and 28,045′ south and longitudes 28,030′ and 28,037′ east (**Figure 1**). With an average yearly rainfall of 780 mm, the region is in the summer rainfall zone, with the wet season lasting from September to April. The upper Karoo Sequence is represented by the rock formations, intruded by dolerite dikes and sills.

The park's vegetation is primarily grassland, with a very small percentage of woodland and forest communities found along rivers and in sheltered rocky areas protected from fire. Mucina and Rutherford [10] identified five dominant vegetation units in the study area: Eastern Free State Sandy Grassland (Gm4), Northern Drakensberg Highland Grassland (Gd5), Drakensberg-Amathole Afromontane Fynbos (Gd6), Lesotho Highland Basalt Grassland (Gd8), and Basotho Montane Shrubland (Gm5).

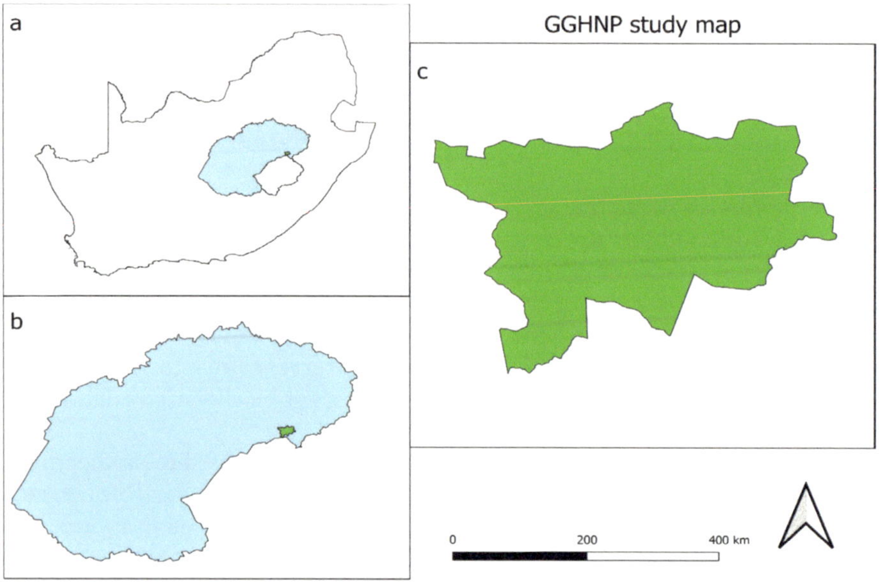


**Figure 1.**
*The study area showing the location of Golden Gate Highlands National Park (c) in the context of the rest of South Africa (a) and the Free State Provinces (b).*

## 2.2 Methodologies

### *2.2.1 Species occurrence*

Digitized natural history collection occurrence data from the Global Biodiversity Information Facility (GBIF) was used to map the historical distribution of the Afrotemperate forests in South Africa Previous studies on the classification and descriptions of vegetation in the GGHNP identified two forest communities, *Olinia emarginata-Podocarpus latifolius* and *Kiggelaria africana* [20, 42]. Three species namely *Olinia emarginata*, *Podocarpus latifolius*, and *Kiggelaria africana* were therefore used as indicator species for the Afromontane forests. Forest and woodland community shape files for GGHNP were sourced from the previous study undertaken by the authors in the GGHNP [20].

### *2.2.2 Bioclimatic variables*

Nineteen historical bioclimatic variables with 2.5 minutes (approximately 5 km) spatial resolution were sourced from WorldClim v2.1 in DIVA-GIS 7.5 software (http://www.diva-gis.org/climate.htm). The bioclimatic variables represent annual trends (e.g., mean annual temperature and annual precipitation), seasonality (e.g., annual range in temperature and precipitation), and extreme or limiting environmental factors (e.g., temperature of the coldest and warmest month, and precipitation of the wet and dry quarters) [30].

The variables are coded as follows: annual mean temperature (Bio 1, °C), mean diurnal range (Bio 2, °C), isothermality (Bio2/Bio7) (×100) (Bio 3), temperature seasonality (standard deviation *100) (Bio 4, Coeff. of variation), the maximum temperature of the warmest month (Bio 5, °C), minimum temperature of the coldest month (Bio 6, °C), temperature annual range (Bio 7, °C), mean temperature of the driest quarter (Bio 9, °C), mean temperature of warmest quarter (Bio 10, °C), mean temperature of coldest quarter (Bio 11, °C), annual precipitation (Bio 12, mm), precipitation of wettest month (Bio 13, mm), precipitation of driest month (Bio 14, mm), precipitation seasonality (coefficient of variation) (Bio 15, Coeff. Of variation), precipitation of wettest quarter (Bio 16, mm), precipitation of driest quarter (Bio 17, mm), precipitation of warmest quarter (Bio 18, mm) and precipitation of coldest quarter (Bio19, mm).

#### *2.2.3 Species distribution models*

*Olinia emarginata-Podocarpus latifolius* and *Kiggelaria africana* forests were the two dominant forest communities in the GGNHNP, according to earlier studies on vegetation classification and descriptions [20, 42]. The Global Biodiversity Information Facility (GBIF; https://www.gbif.org/) provided us with the geographical distribution records for *Olinia emarginata*, *Podocarpus latifolius*, and *Kiggelaria africana* in South Africa. The authors' earlier study served as the source of the forest and woodland community shape files for GGHNP [20]. After that, samples that were duplicates or lacked accurate geographic coordinates were eliminated from the national geographical distribution records. Lastly, Google Maps was used to confirm the longitude and latitude of the few samples with exact coordinate data (http://www.google.cn/maps).

We employed widely used models for species distribution, such as Random Forest [40] in the R statistical environment [43] and the Bioclim analysis tool in the DIVA-GIS program [32]. In studies of niche dynamics, recent research has demonstrated that the scale of investigation is crucial, whether spatial, temporal, environmental, or phylogenetic [44]. The spatial resolution at which climatic or species distribution data are available can be linked to the effects of scale (e.g., in studying the climatic niche) [44]. Because data availability, including the availability of climate observations and future climate simulations [45, 46], is greater at a larger scale that becomes relevant to larger extents [47], the SDM was conducted in this study using a national scale. At the local scale, we used GGHNP as a study area and overlayed the forest and woodland communities' shape files [20] on the national projections.

#### *2.2.4 Model validation*

The species distribution data was divided into two parts to validate the SDM used: 75% of the data was used for calibration, and the remaining 25% was used for testing and validation 10 times [23]. The chosen species' locality points were regarded as belonging to a single class. The outcome was saved as a grid-formatted output file. The area under the receiving operator curve, or ROC, was then constructed using a stack data set [23, 48]. The accuracy of the models employed was assessed using the area under the curve (AUC) of a receiver operator characteristic (ROC) [48]. AUC values fall between 0 and 1, with AUC < 0.5 denoting models with lower performance, 0.8–0.9 being good, and 0.9–1 being excellent [48].

#### *2.2.5 Principal components analysis (PCA)*

For principal component analyses (PCA), 19 bioclimatic variables with varying measurement units were standardized to a mean of 0 and a standard deviation of 1; data transformation was carried out using the PAST 4.3 program [49, 50]. PCA was also used to identify the key variables influencing the spatial distribution of Afrotemperate forest species [51, 52]. Following PCA screening, the most dominant climatic factors were chosen for variable statistics based on the rank of their factor loadings [51, 52]. Random Forest was also used to assess the most important variables influencing the occurrence and distribution of the Afrotemperate forests in the South African landscape.

## 3. Results

### 3.1 Species distribution model

The SDM divides habitat suitability into six categories: not suitable (0%), low (0–2.5%), medium (2.5–5.5%), high (5–10%), very high (10–20%), and excellent (20–46%). The SDM results revealed that the most suitable habitat for the Afrotemperate forest species is along the Drakensberg Mountain range, which runs from Limpopo Province in the northeast to Western Cape Province in the southwest (**Figure 2**). In South Africa's Southern Cape, the Knysna-Tsitsikamma forest complex and KwaZulu-Midlands in Kwazulu-Natal Province are dominated by Afrotemperate forests. The Maluti Drakensberg Transfrontier Project (MDTP) also conserves a significant portion of the Afrotemperate forest [10, 53], covers approximately 28,000 $km^2$ and includes the provinces of KwaZulu-Natal, the Free State, and the Eastern Cape along the border between South Africa and the Kingdom of Lesotho (**Figure 2**).

The digital elevation model for the GGHNP (**Figure 3**) shows that Afrotemperate forests and woodland communities exist between 1277 and 2375 m above sea level. Lower-lying areas in the eastern and northern sections of the GGHNP were found to be highly suitable (5–10%) for woodland species such as *Euclea* and *Protea* that grow

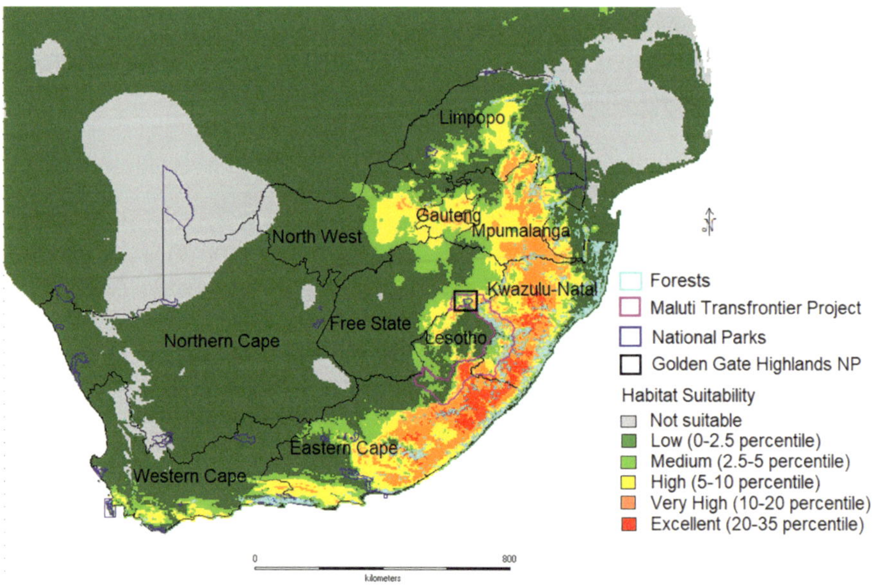


**Figure 2.**
*Species distribution model showing suitable habitats for the distribution of the Afrotemperate forests in South Africa. The light blue polygons represent the forest Biome [10].*

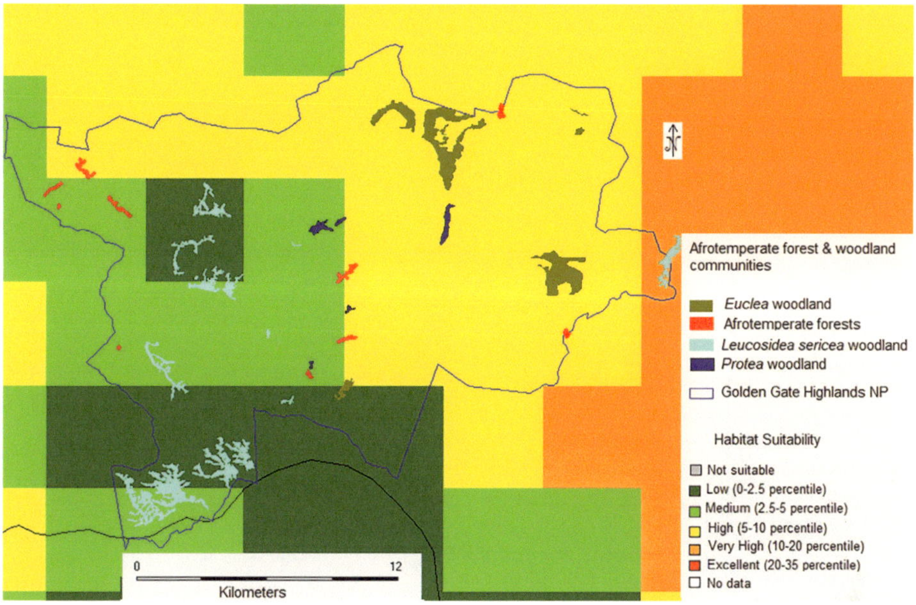


**Figure 3.**
*Species distribution model showing suitable habitats for the distribution of the Afrotemperate forests in the Golden Gate Highland National Park.*

in open grassland. *Euclea* woodland grows at the base of cliffs, under large boulders, and in small clumps in rocky open grassland. The *Protea* woodland community grows in open grassland and is associated with well-drained soils on the foothills and midslopes [20]. The south and west sections of the park were mostly suitable (2.5–5%) for the Afrotemperate forest species such as *Podocarpus latifolius*, *Kiggelaria africana*, and *Olinia emarginata*, occurring in protected gorges. The *Leucosidea sericea* woodland had low habitat suitability (0–2.5%) in the midslope but medium suitability (2.5–5%) in the riverine areas.

### 3.2 Important climatic variables

The results from the principal components analysis (PCA) analysis showed that the bioclimatic data was accounted for by the first four principal components (**Figure 4**). The loading values from PCA are equal to the coefficients of the variables and provide information about which variables give the largest contribution to the components. The results from the PCA analysis showed the four principal components to have the highest eigenvalue, explaining 91.62% of the variance.

The first principal components explained 34.9% of the variation, and the four main bioclimatic factors with high loading values included Bio6, Bio14, Bio17, and Bio19. The above bioclimatic variables showed the influence of extreme environmental factors (cold and dry conditions) and the variation between the coldest and hottest months.

The second axis explained 21.55% of the variation, and the four main bioclimatic factors with high loading values included Bio1, Bio5, Bio10, and Bio11. The above reflects the response of the forest species to annual trends, seasonality, and tolerance to extreme temperatures. The third axis explained 21.17% of the variation, and the

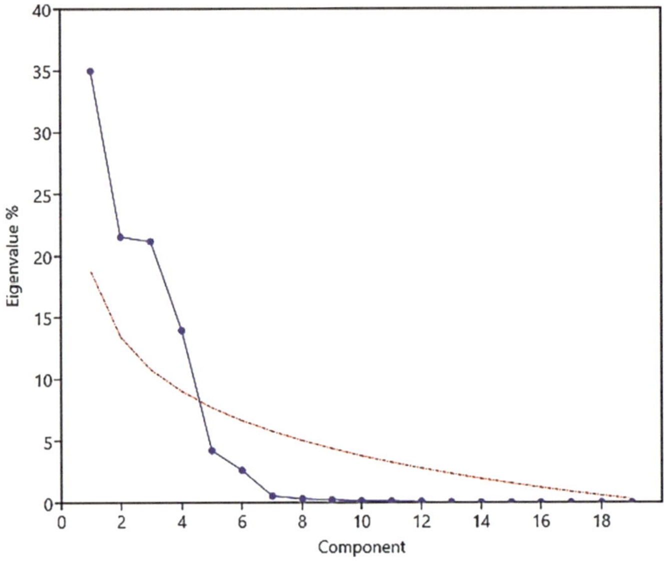


**Figure 4.**
*The scree plot showing eigenvalues and the number of significant principal components of the Afrotemperate forest species for 19 bioclimatic variables in the Golden Gate Highlands National Park.*

four main bioclimatic factors with high loading values were Bio12, Bio13, Bio16, and Bio18. Again, precipitation was of the greatest significance for the spatial distribution of the Afrotemperate forest species. The fourth axis explained 13.93% of the variation, and the four main bioclimatic factors with high loading values included Bio4, Bio9, Bio15, and Bio19.

Based on the rank of the factor loadings for the first four principal components, PCA analysis showed that the four most important variables influencing spatial distribution of the Afrotemperate forest are Bio1, Bio12, Bio13, and Bio16 (**Figure 5**).

The Afrotemperate forest occurred in areas with relatively high rainfall frequencies of 430–860 mm per annum (**Figure 6**). The highest frequency for precipitation of the wettest month and quarter were between 87.4–174.8 and 225.6–451.2 mm respectively. The highest frequency of annual mean temperature was between 12.9°C and 18.2°C. This shows that Afrotemperate forest species occur in areas with seasonal temperature variation and are largely confined to higher rainfall areas or where soil moisture is not limited for extended periods.

Random Forests analysis showed that the four most important variables influencing the spatial distribution of the Afrotemperate forest are Bio5, Bio12, Bio13, Bio18, and Bio19, all increasing the mean square error by more than 10% when randomly permuted (**Figure 7**). %IncMSE in Random Forest indicates the increase of the Mean Squared Error when the given variable is randomly permuted.

### 3.3 Model validation

The AUC values of more than 0.98 for all bioclimatic variables indicated that both the Bioclim and Random Forest models were excellent at predicting species distribution, with the latter performing slightly better (**Figure 8**). Kappa statistics were used to test the accuracy of prediction for Afrotemperate forest species. Bioclim also returned a kappa value of 0.885, indicating that the model and test data were nearly identical.

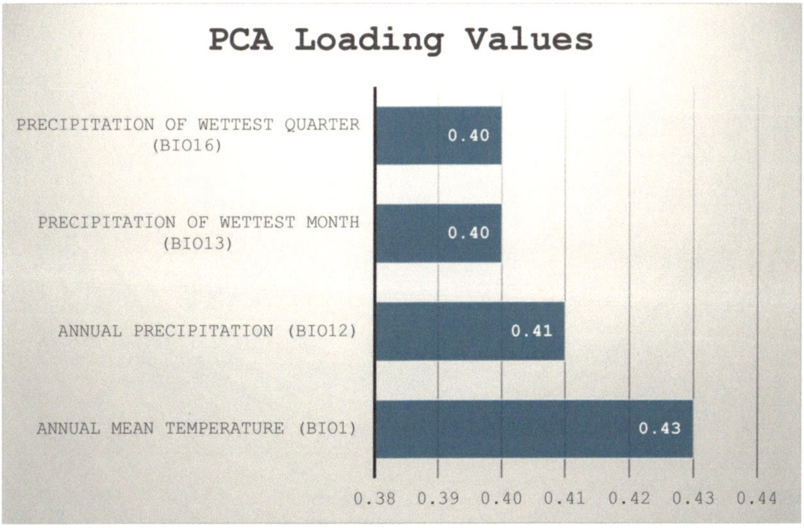


**Figure 5.**
*Principal components analysis for the four most influential bioclimatic variables influencing the spatial distribution of the Afrotemperate forests in South Africa.*

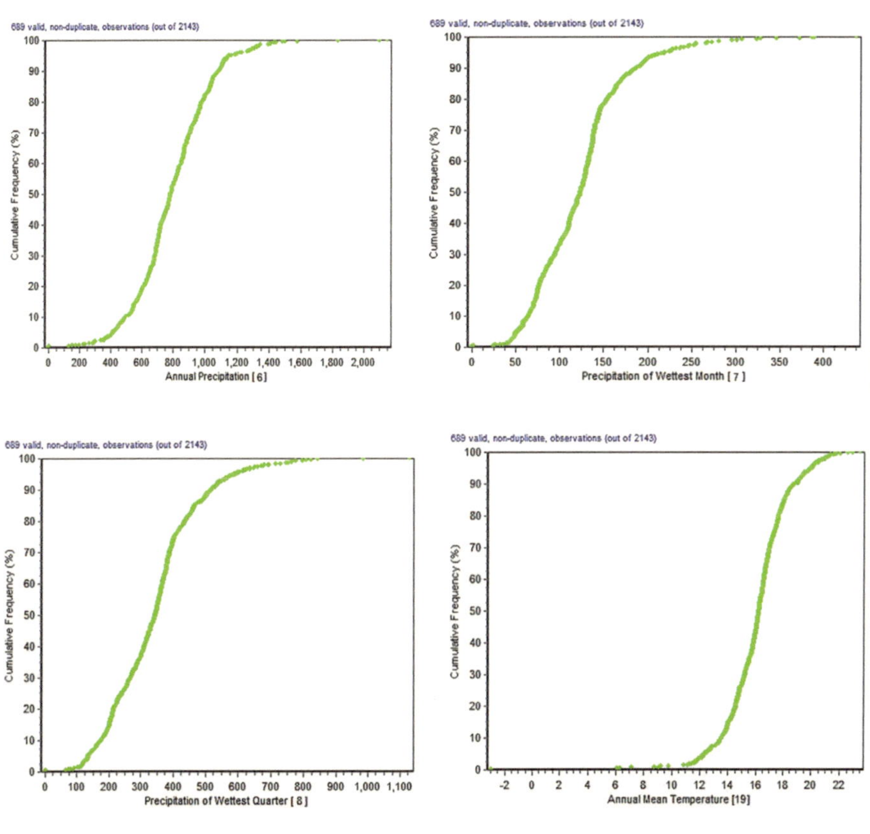


**Figure 6.**
*Frequencies of dominant climatic factors, annual mean temperature; annual precipitation; precipitation of the wettest quarter, and precipitation of the wettest month associated with the Afrotemperate forests in South Africa.*

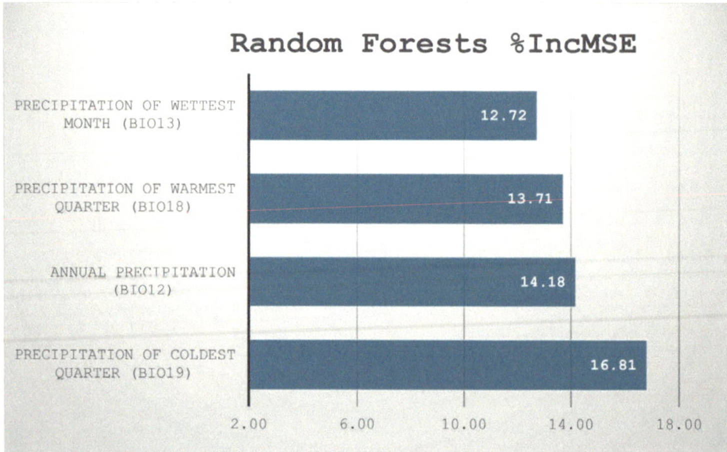


**Figure 7.**
*Random Forests (RF) analysis showing the four most important variables influencing the spatial distribution of the Afrotemperate forest in South Africa.*

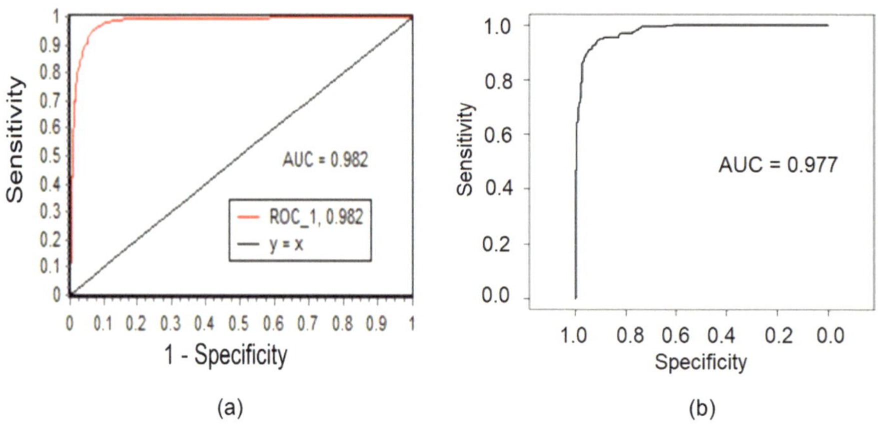


**Figure 8.**
*Models' validation performance for the spatial distribution of the Afromontane forest species. (a) Bioclim model and (b) Random Forest.*

## 4. Discussion

According to the findings of SDMs, water requirements and temperature are the most important bioclimatic variables influencing habitat suitability for Afrotemperate forest species. These included the annual mean temperature, annual precipitation, precipitation of the wettest month, precipitation of the wettest quarter, precipitation of the warmest quarter, and precipitation of the coldest quarter. The findings support the fact that forests persist in areas with a mean annual rainfall of more than 525 mm during the winter and more than 725 mm during the summer [10, 54, 55]. The Afrotemperate forest's distribution range along the Drakensberg range also contributes to its persistence at various altitudes with both summer and winter rainfall.

Previous research discovered that as one moves from the lower slopes to the summit, the vegetation decreases in stature, with tall forests giving way to scrub forests and thickets [56]. This is also supported by the digital elevation model findings in the current study, which show that increasing altitude is associated with low to medium-suitability habitats for Afrotemperate forests in the GGHNP. Elevation in the GGHNP was identified as one of the environmental factors limiting the distribution of woodland communities [20]. Forests and woodland communities were absent above 2757 m and below 1662 m above sea level [20]. In the Natal Drakensberg, the woodland community limit lies in the montane and subalpine vegetation belts just below 2500 m above sea level [57, 58].

Nationally, the Afrotemperate forests currently dominate the landscapes in only a few areas, including the Knysna-Tsitsikamma forest complex in South Africa's Southern Cape and mistbelt forests in the KwaZulu-Midlands. Drakensberg montane forests cover approximately 7025 ha in total, all of which are located within the Maluti Drakensberg Transfrontier Project (MDTP) region [59]. Eastern mistbelt forests, on the other hand, cover approximately 33,000–35,000 ha and span a wide latitudinal range from the KwaZulu-Natal midlands to Mt. Ayliff in the Eastern Cape [59]. The MDTP Bioregion contains approximately a quarter of the Eastern mistbelt forest's total area (25.6%; 8708 ha). Eastern mistbelt forests grow at medium altitudes (850–1600 m) and are primarily found on south and southeast-facing slopes [59].

In terms of model accuracy, the AUC values of 0.982 and 0.977 for Bioclim and Random Forest respectively indicated an excellent model for predicting species distribution on a national scale. Bioclim was one of the most popular SDM packages, and it was regarded as one of only three "well-established modeling methods" [22, 31]. Compared to other models, Bioclim provides a straightforward approach to species distribution modeling and is a highly flexible and powerful tool for evaluating distributions on a variety of spatial scales with small sample sizes [31, 60]. However, it is important to note that even though the model shows the potential areas of spatial change of Afrotemperate, especially in the Free State Province (GGHNP), the availability of the areas would depend on the anthropogenic factors, multiple land use practices (i.e., agriculture and mining), and other unpredictable natural phenomena that may negatively impact on the distribution of these forests.

## 5. Conclusion

The GGHNP's Afrotemperate forests serve as important biogeographical links between larger forest areas in southern Africa, providing habitat for a diverse range of forest species. Even though they cover a small area and have a much lower canopy height than the southern Cape and Tsitsikamma Afrotemperate forests, they continue to provide floristic diversity and a home for South African endemic tree species such as *Olinea emarginata* and *Scolopia mundii*.

The SDMs model proved to be effective in projecting the current geographic ranges of Afrotemperate forest species. The statistical models validated the current spatial distribution of forest species. The high AUC values (0.982 and 0.977) and kappa statistics (0.81–0.99) demonstrated that Bioclim and RF were excellent models for predicting species distribution, and the accuracy of prediction of Afrotemperate forest species indicated a near-perfect agreement between the model and the test data. Water requirements and temperature were discovered to be the most important bioclimatic factors influencing habitat suitability for Afrotemperate forest species.

A significant future increase in habitat suitability for Afrotemperate forests in the Free State Province is associated with rising temperatures, which results in fewer frost events. The current densification and encroachment of species such as *Leucosidea sericea*, as well as the emergence of *Euclea* woodland in the GGHNP's lower-lying areas, demonstrate the influence of bioclimatic variables on the current and future spatial distribution of Afrotemperate forest species. However, on a national scale, Afrotemperate forests will continue to dominate landscapes in areas like the Knysna-Tsitsikamma forest complex and KwaZulu-Midlands.

Understanding the effects of climate on biodiversity, as well as the various levels of response to climatic variation, is critical for prioritizing biodiversity conservation efforts. The findings of this study will help to conserve Afrotemperate forests and make important predictions about future suitable habitats in the face of climate change.

## Author details

Mahlomola Ernest Daemane[1*], Samuel Adelabu[2] and Abel Ramoelo[3]

1 Conservation Services Department, South African National Parks, Kimberley, South Africa

2 Department of Geography, University of the Free State, Bloemfontein, South Africa

3 Centre for Environmental Studies, Department of Geography, Geoinformatics and Meteorology, University of Pretoria, South Africa

*Address all correspondence to: ernest.daemane@sanparks.org

## References

[1] Calabrese V, Carranza ML, Evangelista A, Marchetti M, Stinca A, Stanisci A. Long-term changes in the composition, ecology, and structure of Pinus mugo scrubs in the Apennines (Italy). Diversity. 2018;**10**(3):70. DOI: 10.3390/d10030070

[2] Myers N, Mittermeier RA, Mittermeier CG, da Fonseca GAB, Kent J. Biodiversity hotspots for conservation priorities. Nature. 2000;**403**(6772):853-858. DOI: 10.1038/35002501

[3] Pauli H, Gottfried M, Dirnböck T, Dullinger S, Grabherr G. Assessing the long-term dynamics of endemic plants at summit habitats. In: Mediterranean Type Ecosystem. Vol. 167. Berlin, Heidelberg: Springer; 2003. pp. 195-207. DOI: 10.1007/978-3-642-18967-8_9

[4] Winkler M, Lamprecht A, Steinbauer K, Hülber K, Theurillat JP, Breiner F, et al. The rich sides of mountain summits—A pan-European view on aspect preferences of alpine plants. Journal of Biogeography. 2016;**43**(11):2261-2273. DOI: 10.1111/jbi.12835

[5] Nogués-Bravo D, Araújo MB, Romdal TS, Rahbek C. Scale effects and human impact on the elevational species richness gradients. Nature. 2008;**453**(7192):216-219. DOI: 10.1038/nature06812

[6] Carlson BZ, Randin CF, Boulangeat I, Lavergne S, Thuiller W, Choler P. Working toward integrated models of alpine plant distribution. Alpine Botany. 2013;**123**(2):41-53. DOI: 10.1007/s00035-013-0117-4

[7] Chapman HM, Olson SM, Trumm D. An assessment of changes in the montane forests of Taraba State, Nigeria, over the past 30 years. Oryx. 2004;**38**(3):282-290. DOI: 10.1017/S0030605304000511

[8] Cordeiro NJ, Burgess ND, Dovie DB, Kaplin BA, Plumptre AJ, Marrs R. Conservation in areas of high population density in sub-Saharan Africa. Biological Conservation. 2007;**134**(2):155-163. DOI: 10.1016/j.biocon.2006.08.023

[9] Brown LR, du Preez PJ. Alpine vegetation of temperate mountains of southern Africa. Encyclopedia of the World's Biomes. 2019;**2020**: 395-404. DOI: 10.1016/B978-0-12-409548-9.11892-5

[10] Mucina L, Rutherford MC, editors. The Vegetation of South Africa, Lesotho and Swaziland, Strelitzia 19. Pretoria: South African National Biodiversity Institute; 2006

[11] Alvarez G, Salas EAL, Harings NM, Boykin KG. Projections of future suitable bioclimatic conditions of parthenogenetic whiptails. Climate. 2017;**5**:34. DOI: 10.3390/cli5020034

[12] Sexton JP, Mcintyre PJ, Angert AL, Rice KJ. Evolution and ecology of species range limits. Annual Review of Ecology, Evolution, and Systematics. 2009;**40**:415-436. DOI: 10.1146/annurev.ecolsys.110308.120317

[13] Pecchi M, Marchi M, Burton V, FrancescaGiannetti F, Moriondo M, Bernetti J, et al. Species distribution modelling to support forest management: A literature review. Ecological Modelling. 2019;**411**:108817. DOI: 10.1016/j.ecolmodel.2019.108817

[14] Hill TR. Description, classification and ordination of the dominant

vegetation communities, Cathedral Peak, KwaZulu-Natal Drakensberg. South African Journal of Botany. 1996;**62**(5):263-269

[15] Linder HP. The evolution of African plant diversity. Frontiers in Ecology and Evolution. 2014;2:38. DOI: 10.3389/fevo.2014.00038

[16] Erschbamer B, Kiebacher T, Mallaun M, Unterluggauer P. Short-term signals of climate change along an altitudinal gradient in the South Alps. Plant Ecology. 2009;**202**(1):79-89. DOI: 10.1007/s11258-008-9556-1

[17] Lézine AM, Izumi K, Kageyama M, Achoundong GA. 90,000-year record of Afromontane Forest responses to climate change. Science. 2019;**363**(6423):177-181. DOI: 10.1126/science.aav6821

[18] Meadows ME, Linder HP. A palaeoecological perspective on the origin of Afromontane grasslands. Journal of Biogeography. 1993;**20**(4):345-355. DOI: 10.2307/2845584

[19] Adie H, Kotze DJ, Lawes MJ. Small fire refugia in the grassy matrix and the persistence of Afrotemperate forest in the Drakensberg Mountains. Scientific Reports. 2017;7(1):6549. DOI: 10.1038/s41598-017-06747-2

[20] Daemane ME, Ramoelo A, Adelabu S. The spatial distribution of the woodland communities and their associated environmental drivers in the Golden Gate Highlands National Park, South Africa. Koedoe. 2021;**63**(1):a1672. DOI: 10.4102/koedoe.v63i1.1672

[21] Elith J, Leathwick JR. Species distribution models: Ecological explanation and prediction across space and time. Annual Review of Ecology, Evolution, and Systematics. 2009;**40**:677-697. DOI: 10.1146/annurev.ecolsys.110308.120159

[22] Elith J, Graham CH, Anderson RP, Dudík M, Ferrier S, Guisan A, et al. Novel methods improve prediction of species distributions from occurrence data. Ecography. 2006;**29**(2):129-151. DOI: 10.1111/j.2006.0906-7590.04596.x

[23] Xie C, Yu X, Liu D, Fang Y. Modelling suitable habitat and ecological characteristics of old trees using DIVA-GIS in Anhui Province, China. Polish Journal of Environmental Studies. 2020, 2020;**29**(2):1931-1943. DOI: 10.15244/pjoes/110346

[24] Austin M. Species distribution models and ecological theory: A critical assessment and some possible new approaches. Ecological Modelling. 2007;**200**(1):1-19. DOI: 10.1016/j.ecolmodel.2006.07.005

[25] Loiselle BA, Howell CA, Graham C, Goerck JM, Brooks T, Smith KG, et al. Avoiding pitfalls of using species distribution models in conservation planning. Conservation Biology. 2003;**17**(6):1591-1600. DOI: 10.1111/j.1523-1739.2003.00233.x

[26] Pearson RG, Thuiller W, Arau MB, Mcclean C, Miles L. Model-based uncertainty in species range prediction. Journal of Biogeography. 2006;**33**(10):1704-1711. DOI: 10.1111/j.1365-2699.2006.01460.x

[27] Segurado P, Araujo MB. An evaluation of methods for modelling species distributions. Journal of Biogeography. 2004;**31**(10):1555-1568. DOI: 10.1111/j.1365-2699.2004.01076x

[28] Seoane J, Carrascal LM, Alonso L, Palomino D. Species-specific traits associated to prediction errors in bird habitat suitability modelling. Ecological

Modelling. 2005;**185**:299, 2005-308. DOI: 10.1016/j.ecolmodel.2004.12.012

[29] Mapuru MJ, Sifiso Xulu S, Gebreslasie M, Daemane ME. Exploring environmental factors that influence the distribution of poplar trees. African Journal of Ecology. 2024;**62**(4):e13310. DOI: 10.1111/aje.13310

[30] Fick SE, Hijmans RJ. WorldClim 2: New 1km spatial resolution climate surfaces for global land areas. International Journal of Climatology. 2017;**37**(12):4302-4315

[31] Booth TH, Nix HA, Busby JR, Michael F. BIOCLIM: The first species distribution modelling package, its early applications and relevance to most current Maxent studies. Diversity and Distributions. 2014;**20**(1):1-9. DOI: 10.1111/ddi.12144

[32] Busby JR. BIOCLIM—A bioclimate analysis and prediction system. Plant Protection Quarterly. 1991, 1991;**1.6**(1):8-9

[33] Lawrence RL, Wood SD, Sheley RL. Mapping invasive plants using hyperspectral imagery and Breiman Cutler classifications (randomForest). Remote Sensing of Environment. 2006;**100**:356-362

[34] Falkowski MJ, Evans JS, Martinuzzi S, Gessler PE, Hudak AT. Characterizing forest succession with lidar data: An evaluation for the inland Northwest, USA. Remote Sensing of Environment. 2009;**113**:946-956

[35] Cutler DR, Edwards TC, Beard KH, Cutler A, Hess KT, Gibson J, et al. Random Forests for classification in ecology. Ecology. 2007;**88**:2783-2792

[36] Evans JS, Cushman SA. Gradient modeling of conifer species using Random Forests. Landscape Ecology. 2009;**24**:673-683

[37] Murphy MA, Evans JS, Storfer AS. Quantifying *Bufo boreas* connectivity in Yellowstone National Park with landscape genetics. Ecology. 2010;**91**:252-261

[38] Prasad AM, Iverson LR, Liaw A. Newer classification and regression tree techniques: Bagging and Random Forests for ecological prediction. Ecosystems. 2006;**9**:181-199

[39] Rehfeldt GE, Crookston NL, Warwell MV, Evans JS. Empirical analyses of plant-climate relationships for the western United States. International Journal of Plant Sciences. 2006;**167**:1123-1150

[40] Breiman L. Statistical modeling: The two cultures. Statistical Science. 2001;**16**:199-231

[41] Hastie T, Tibshirani R, Friedman J. The Elements of Statistical Learning: Data Mining, Inference, and Prediction. 2nd ed. New York: Springer; 2009

[42] Manfred DM. The structure and floristic composition of Afromontane forests in the Golden Gate Highlands National Park [B.Sc. Honours thesis]. Johannesburg: Department of Botany, University of Johannesburg; 1990

[43] Buckley LB, Jetz W. Environmental and historical constraints on global patterns of amphibian richness. Proceedings of the Royal Society. Biological Sciences. 2007;**274**:1167, 2007-1173. DOI: 10.1098/rspb.2006.0436

[44] Pearman PB, Guisan A, Broennimann O, Randin C. Niche dynamics in space and time. Trends in Ecology & Evolution. 2008;**23**(3):149-158. DOI: 10.1016/j.tree.2007.11.005

[45] Austin MP, Van Niel KP. Improving species distribution models for climate change studies: Variable selection and scale. Journal of Biogeography. 2011;**38**:1-8

[46] Manzoon SA, Geoffrey G, Martin L. Species distribution model transferability and model grain size–finer may not always be better. Scientific Reports. 2018;**8**(1):7168

[47] Connor T, Viña A, Winkler JA, et al. Interactive spatial scale effects on species distribution modelling: The case of the giant panda. Scientific Reports. 2019;**9**:14563

[48] Hand DJ. Measuring classifier performance: A coherent alternative to the area under the ROC curve. Machine Learning. 2009;**77**:103-123. DOI: 10.1007/s10994-009-5119-5

[49] Hammer O, Harper DAT, Ryan PD. PAST: Paleontological statistics software package for education and data analysis. Paleontologia Electronica. 2001;**4**(1). 9 p

[50] Jackson FA. Stopping rules in principal components analysis: A comparison of heuristical and statistical approaches. Ecology. 1993;**74**(8):2204-2214. DOI: 10.2307/1939574

[51] Davis JC. Chapter 4: Clarification. In: Statistics and Data Analysis in Geology. 3rd ed. New York: John Wiley & Sons Inc.; 1986. pp. 238-239

[52] Krebs C. Ecology: The Experimental Analysis of Distribution and Abundance. 6th ed. Pearson Education Limited; 2014. pp. 378-386

[53] Lawes MJ, Adie H, Eeley H, Kotze DJ, Wethered R. An assessment of the forests of the maloti-Drakensberg transfrontier bioregion, with reference to important ecosystem processes [thesis]. Forest Biodiversity Research Unit, School of Biological and Conservation Sciences. South Africa: University of KwaZulu-Natal; 2007

[54] Geldenhuys CJ. Distribution, size and ownership of forests in the southern cape. Southern African Forestry Journal. 1991;**158**(1):51-66

[55] Rutherford MC, Westfall RH. Biomes of Southern Africa: An Objective Categorization, Memoirs of the Botanical Survey of South Africa. Vol. 54. Pretoria: National Botanical Institute; 1994. pp. 1-98

[56] White F. The afromontane region. In: Werger MJA, editor. Biogeography and Ecology of Southern Africa. Berlin/Heidelberg: Springer; 1978. pp. 463-513. DOI: 10.1007/978-94-009-9951-0_11

[57] Chawla A, Rajkumar S, Singh K, Lal B, Singh R, Thukral A. Plant species diversity along an altitudinal gradient of Bhabha Valley in western Himalaya. Journal of Mountain Science. 2008;**5**:157-177. DOI: 10.1007/s11629-008-0079-y

[58] Killick DJB. An Account of the Plant Ecology of the Cathedral Peak Area of the Natal Drakensberg, Memoirs of the Botanical Survey of South Africa. Vol. 341963. Printed in the Republic of South Africa by the Government Printer, Pretoria. pp. 1-178

[59] Mucina L, Lötter MC, Rutherford MC, van Niekerk A, Macintyre PD, Tsakalos JL, et al. Forest biomes of Southern Africa. New Zealand Journal of Botany. 2021;**60**(4):1-52. DOI: 10.1080/0028825X.2021.1960383

[60] Booth TH, Jovanovic T. Climatology of *Acacia mearnsii*. 1. Characteristics of natural sites and exotic plantations. NewForests. 1988;**2**:17-30. DOI: 10.1007/BF00039039

Section 4

# Sustainable Use of Forest Resources and Livelihoods

Chapter 9

# Wild Edible Plants and Their Role in Alleviating Nutritional Deficiency: An Indian Context

*Aakanksha Pandey, Archana Mishra, Smaranika Mohanta, Abha Manohar Kandileri, Gopal Shukla and Sumit Chakravarty*

**Abstract**

World Health Organization estimates that nutrient deficiency affects more than 2 billion people globally which can only be defeated by a well-balanced diet. Awareness about proper nutrition, nutritive value of food, and balanced dieting practices are the prerequisites toward adopting healthy eating habits. In many developing countries, including India, rural and tribal populations rely on wild edible plants to meet their daily nutritional needs due to insufficient access to adequate food. Wild edible fruits and vegetables require less maintenance than conventionally grown plants and are rich in dietary fibers, micronutrients, vitamins, and mineral compounds essential for health and immunity. There is a need to diversify the diet including traditionally acceptable wild edible plants through scientific interventions. Hence, specific attention must be paid to developing different healthcare programs to encourage a healthy diet and lifestyle, by incorporating the lesser-known edible wild plants to reduce nutritional deficiency and thereby enhance food security.

**Keywords:** food security, medicinal value, micronutrients, nutrition, traditional knowledge, UN-SDGs

## 1. Introduction

Nutritional deficiencies have become a serious global problem, especially in areas, where the diet lacks variety. There are approximately 40 vitamins and mineral compounds that are considered essential for physical and mental development, the immune system, and metabolic processes [1]. A vast range of nutrients is required for humans to lead a healthy life. Nutrients play a significant part in the proper growth and development of the human body, and their deficiency impacts health leading to low immunity, low productivity, and underdevelopment [2]. Globally the food habits of humans have changed manifold, and it lacks major micro-nutrients. Wild plants have been recognized to have the potential to meet household food and income security of the estimated 2800 species of vascular plants of Orissa state (India), and about 150 wild edible fruit species occurring in different parts of eastern India's deciduous forests are consumed in various quantities by rural communities [3]. Over the decades as civilizations grow traditional plants that were a part of the human diets (food, livelihood, medicine, income generation, cultural, and religious purposes) have been replaced by nearly thirty domesticated species which are less nutritive and "modern" ones (three principal cereal grains, viz. rice, wheat, and maize), which has caused serious nutritional deficiencies [4, 5]. Wild edible flora (food and medicine) are plant resources that grow in natural conditions (wild) and are harvested or collected for human consumption and used as food, dietary supplements, and medical treatment. A diverse range of wild uncultivated plants and their parts (e.g., leafy shoots, fruits, seeds, underground organs, and flowers) are still being consumed regularly and complement human adaptability and a variety of human gastronomic choices. They tend to supplement proteins, essential minerals, micronutrients, and vitamins that enrich dietary quality. In other situations, thousands of edible species remained wild or semi-wild and were left out in the course of domestication, and climate change and other human-made casualties have made many species into extinction and lands unsuitable for the particular species to grow and survive [6]. The human race should ensure adequate nutritious food security throughout the world for the betterment of fellow beings belonging to poor and war-ridden countries sustainably by promoting traditional or locally grown foods [7]. A vast range of nutrients is required for humans to lead a healthy life. Nutrients play a significant part in the proper growth and development of the human body, and their deficiency impacts health leading to low immunity, low productivity, and underdevelopment [2].

Wild edible plants provide staple food, and important nutrient and vitamin supplements for indigenous people but are unknown to others and hundreds of millions of people, mostly in underdeveloped and developing countries, derive a substantial part of their subsistence and income from wild plant products [8]. Tibetans in Shangri-la region, Yunnan, China, are using herbs (43.5%) and shrubs (26.8%) as WEP, and in total, vegetable (41.9%) is the most used category followed by fruit (40.8%). In addition to edible use, 71.4% of the reported wild edible plants (120 species) have additional uses. Thirty-one species (18.5%) are also used as medicine, most are herbs (19 species) or trees (6 species), and these medicinal plants are used to treat gastropathy, cough, fever, rheumatism, dysentery, fractures, dyspepsia, hemoptysis, and asthma. The people of Nepal are using WEP cooked as a vegetable (36 species), eaten as fruit (44), prepared as pickle1 (15), and used as a spice (3), and 19 species (24%) were also used as medicine [9]. In India, a total of 60 species of leafy vegetables of wild edible plants were mainly collected from herbs (67%) followed by shrubs (19%), trees (10%), and climbers (5%), and 60% of the wild edible fruits were collected from trees, while 17% climbers, 16% shrubs, and 7% herbs and additionally. The mushrooms (19 wild edible mushrooms) were consumed in diets after cooking as vegetables by adding salt and spices, or occasionally they were pickled to utilize for longer duration [10]. One of the greatest challenges facing the world today is to feed the ever-increasing population. Several countries in the tropics including India suffer from famines and food shortages although they have the potential to produce adequate nutritional bases for their populations [11]. World Health Organization estimates that global nutrient deficiency affects more than 2 billion people which can only be defeated by a well-balanced diet. In 2000, more than 3.7 million deaths could be attributed to underweight children; according to the Food and Agriculture Organization 2004 in addition to these, deficiencies in iron, zinc, or vitamin A caused an additional 750,000 to 850,000 deaths [1]. Awareness about proper nutrition, the nutritive value of food, and eating practices are the first steps toward adopting healthy eating habits. Millions of people in many developing countries or underdeveloped countries do not have enough food to meet their daily nutrient requirements. In the Indian scenario, most rural and tribal inhabitants depend on wild edible plants to meet their food requirements. Wild edible plants are crucial components of diversified diets and resilient food systems from local to global scales. Several nations have a double burden of malnutrition, with under nutrition coexisting with overweight, obesity, and other diet-related disorders [12]. Malnutrition is largely caused by poor-quality meals that are deficient in variety and micronutrients. While less than one billion individuals do not get enough calories [13].

## 2. Relevance of wild edible plants as a source of nutrients

Wild edible plants have always been very important to humans as a source of food, shelter, and many other necessities. Many people all over the world have relied on wild species for food and medicine. WEP are species that are harvested from their natural habitats for human consumption but are not farmed; forests, pastures, and farms are among the surrounding ecosystems. FAO defines them as "plants that grow spontaneously in self-maintaining populations in natural or semi-natural ecosystems and can exist independently of direct human actions". They provide a variety of nutritional and health benefits. They are low-cost, easy to cook, and high in macro and micronutrients. According to the FAO, about one billion people consume wild foods [14]. Daily, approximately one billion people worldwide consume wild foods (mostly from plants). Wild edible plants are an important source of nutrition in many parts of the world, particularly in developing nations such as India. They frequently contribute significantly to the diet, particularly in rural areas. Millions of people in many developing countries lack adequate food to meet their daily needs. Wild edible plants are nutritious and can also supplement nutritional requirements due to their greater nutritional value. Apart from nutritional benefits, the sale of their leaves, fruits, juice, and local drinks can provide income and employment. Most rural people can access affordable medications made from wild plants. Wild edible plants can enhance diets by providing alternative sources of more affordable, nutritionally rich fruits and vegetables that are available all year and can grow in drought-prone, water-stressed areas, and diverse environmental conditions. It is estimated that nearly 75,000 flowering plants out of the millions found around the world are edible. India has a diverse forest cover, which is a major land use after agriculture. India is a biodiversity-rich nation in terms of vegetation and flora, soil status, and environmental conditions that support a diverse range of forest types.

Fruits are the dominant parts of wild edible plants used for medicinal purposes as well as for consumption. Despite providing significant nutritional value, wild edible plants are regarded as a staple diet for low-income families. When compared to people with moderate earnings, the poor and the landless are frequently more reliant on foods from the forest. Food was obtained by ancient societies through hunting and gathering from Mother Nature over time, and some plant species were cultivated in gardens and wild fields by these societies to cater for their various needs [4]. Some forest foods are also consumed in times of scarcity as a substitute for staple foods. In times of famine or natural disaster, wild edible foods can be life-saving. Fruits of trees and shrubs play an important role as emergency food in areas where rainfall is low and erratic. About 150 species of wild plants have been identified in India as potential sources of emergency food. The most important nutritional compounds found in wild plants are carbohydrates in the form of sugars and starch, proteins and lipids in the form of oils, vitamins, minerals, etc. Foods from the forest offer a wide range of nutrients. Different parts of the same species are consumed as food by different population groups. For example, locals consume certain palm varieties' products such as cooked fruits, vegetable oil, palm wine (the sap is high in protein, vitamins, and iron), or flour for baking. Forest fruits are an excellent source of dietary fibers, mineral compounds, vitamins, and a variety of other nutrients that are essential for growth and development as well as human health. The most important nutritional compounds found in wild plants are carbohydrates in the form of sugars and starch, proteins and lipids in the form of oils, vitamins, minerals, etc. [15].

WEP is commonly used in the diets of many communities, especially among indigenous people as a part of their traditional food system. Knowledge of such foods is passed orally from generation to generation in rural communities; wild green plants supply essential nutrients for good health and nutritional security. These include amino acids, vitamins, essential fatty acids, minerals, and dietary fibers. They are only natural sources of folic acid and are also rich in nutrients like iron and calcium. These are the cheapest sources of vitamins and micronutrients in food. For persons who eat purely vegetarian diets, wild greens are the major source of vitamins and micronutrients. In remote rural areas where veggie production is not practiced and market supplies are not organized, residents rely on indigenous green vegetables, both cultivated in kitchen gardens and forests. In comparison with conventionally cultivated species, wild fruits and vegetables require less maintenance and are high in micronutrients. Wild green leafy vegetables enhance hemoglobin content in the body and have good nutritional potential to meet the recommended dietary allowances. Wild edible green leafy plants have a rich nutrient profile and provide a good source of nutrients for the diet to boost nutrition, food security, and livelihoods of the rural population across the country.

## 3. Wild foods as staples in the diet

According to the Food and Agriculture Organization, around one billion people in developing nations rely on wild edible plants for nutrition [16, 17]. Wild edible plants are those that are edible and have not been cultivated but grow like a weed in the main crop, roadside, or in the dense forest, riverside, and lakes that do not require any proper cultivation method. Many rural subsistence households rely on wild plants; it is estimated that at least 1000 million people use them. For example, the leaves of over 100 wild plant species and the fruits of another 200 are consumed in Ghana. More than 220 species of wild plants provide a more significant share of the diet in rural Swaziland than domesticated cultivars. About 150 wild plant species have been identified as emergency food sources in India, Malaysia, and Thailand. In many Indian states, wild food plants have been used as staples.

These wild plants are collected by the rural people and consumed as vegetables, salad, soup, and chutney and can be mixed with many other conventional vegetables [18]. Wild fruits and vegetables contain high sources of minerals, vitamins, proteins, and amino acids. Therefore, these edible wild plants are considered a staple food for indigenous people [14]. Many countries have reported recently an increase in the regular and widespread consumption of wild edible plants, by rural, indigenous communities and particularly in urban counterparts [19]. Wild plants are also rich in many bioactive compounds having antioxidant and therapeutic properties [20].

Poor nutrition is the biggest and most important risk factor for increased infectious disease susceptibility and a key risk factor for various non-communicable disorders. Indigenous tribal societies use wild edible species as food, vegetables, drinks, cooking oils (from seed), spices and sauces, medicine, and industrial goods [21]. Without these wild edible plants, rural communities' cultural, social, and religious belief systems are incomplete. Domestication of these wild food plants has the potential to boost their usage as well as their conservation [16]. Scientific studies have shown that regular eating of veggies and fruits can avoid most cancers on the esophageal, stomach, pancreatic, bladder, and cervical [22]. Research studies highlighted that the edible wild plants are the source of "nutraceuticals", which signifies both nutritional and pharmaceutical values [5, 11]. The wild edible plants are also relatively low in calories, cholesterol-free, and provide unique taste and flavor [23]. Around the world, vegetables are a significant part of the regular dietary component, which plays a vital role in a balanced diet [24]. Conventional and non-conventional green leafy vegetables are excellent sources of minerals that, on regular consumption, overcome the micronutrient deficiency with minimum cost [1, 25]. Moreover, these vegetables also contain both essential and toxic metals in different concentrations and have several toxicological effects on the human body [26] when eaten in excess. From rural Africa, interviews with farmers and herders reported that 67 wild foods, coming from 53 species of plants, are consumed locally and the knowledge of wild foods varied according to ethnic groups and gender. Even in urban areas of Cameroon, wild plant foods (fruits, vegetables, and spices) are available in the cities most of the time, and they spend a considerable part of their food budget on wild foods. An explorative study done in Uganda on wild food plants found 105 species (distributed in 77 genera and 39 families), and most of the wild plants are consumed locally, while some are traded. Difficulty in finding WEP is a reason for them being neglected in the regular diet, and consumption of wild edible plants has been reduced over time. Various countries are now setting up approaches to advance the conservation and sustainable use of wild edible plants [19].

## 4. Nutritional composition of selected wild edible plants

Consumption of fruits and vegetables reduces the risk of cardiovascular disease, stroke, and cancers of the mouth, esophagus, lungs, pharynx, stomach, and colon [14, 27]. Wild edible plants are the cheapest sources of carbohydrates, proteins, amino acids, vitamins (vitamins A, $B_1$, $B_2$, $B_3$, C), essential fatty acids, minerals, and dietary fibers [28, 29]. **Table 1** illustrates the nutritional composition of certain wild edible plants in India. They are natural sources of folic acid and also rich in nutrients like iron and calcium which can be used as the best alternative to the conventional human diet as only plant-based foods are gaining popularity [14, 63]. Wild plants like *Alocacia indica, Asparagus officinalis, Chlorophytum comosum, Cordia myxa, Eulophia ochreata, Momordica dioicia, Portulaca oleracia, Solanum indicum,* etc., are widely consumed in many regions of India [14] as fruits and vegetables. From **Table 1**, it is seen that *Portulaca oleracia* and *Asparagus officinalis* have high amounts of proteins, fats, and calorie values, and these plants are recommended for consumers as vegetables in their diet. The nutritive value of wild medicinal plants, viz. *Eulophia ochrieata, Adiantum philipense, Mucuna prurience,* and *Mucuna championii* for starch, protein, vitamin C, etc., and a maximum of 36.99 and 47.16 mg/g of starch and protein content, respectively, found in *Mucuna Championii. Polygonum runcinatum, Pilea bracteosa,*

| S. No. | Name of Species | Moisture | Energy | Protein | Fiber | Fat | Carbohydrate | Ca | Fe | Mg | P | K | Na | Zn | Cu | Mn | Vit C | B1 | B2 | Vit E | References |
|---|---|---|---|---|---|---|---|---|---|---|---|---|---|---|---|---|---|---|---|---|---|
| 1 | *Aegle marmelos (L.) Corrêa* | 61.00% | — | 1.8 g | 2.90 g | 0.4 g | 31.8 g | 85 mg | 0.6 mg | — | 31.8 mg | 0.6 mg | — | — | — | — | 8000 mg | — | 1.2 mg | — | Omoyeni and Adeyeye [30] |
| 2 | *Aerva lanata (L.) Juss.* | — | — | 22.6 g | — | — | 26.6 g | 51.7 mg | 11 mg | 41.5 mg | 187 mg | 39.4 mg | 10.4 mg | 44.7 mg | — | 1.04 mg | 19 mg | — | — | — | Mishra et al. [31] |
| 3 | *Aesculus indica (Wall. ex Cambess.) Hook.* | — | — | — | — | — | — | 8.2 mg | 8.5 mg | — | 19 mg | 81 mg | — | 705.90 mg | 0.60 mg | 0.50 mg | — | — | — | — | Muszynska et al. [32] |
| 4 | *Agaricus sp (Chiple)* | — | — | — | — | — | 4.50 g | 460 mg | 200 mg | 1150.5 mg | 9690 mg | 3500 mg | 0.05 mg | 54.81 mg | 25 mg | — | 17 mg | 0.6 mg | 5.1 mg | — | Ray et al. [4] |
| 5 | *Agaricus sp. (Patpate)* | — | — | — | — | — | — | 1.53 mg | 0.11 mg | 0.34 mg | 0.76 mg | 2.17 mg | 0.06 mg | 0.6 mg | — | — | — | — | — | — | Ray et al. [4] |
| 6 | *Alternanthera acaulis Andersson* | — | — | — | — | — | — | 50 mg | 1.63 mg | — | — | — | — | — | — | — | 17 mg | — | 0.14 mg | 2 | Leung et al., [33] |
| 7 | *Amaranthus sp.* | 84% | 42 kcal | 4.6 g | 1.8 g | 0.2 g | 8.3 g | 410 mg | 8.9 mg | | 103 mg | — | — | — | — | — | | | | | Soriano et al. [34] |
| 8 | *Amaranthus spinosus L.* | — | — | — | — | — | — | 159 mg | 7.61 mg | 248 mg | 557 mg | 508 mg | 4 mg | 2.87 mg | — | — | 4.20 mg | 0.12 mg | 0.20 mg | 2 IU | Ray et al. [4] |
| 9 | *Amaranthus tricolor L.* | — | — | — | — | — | — | 397 mg | 3.49 mg | — | — | — | 230 mg | 0.18 mg | — | — | 99 mg | — | 0.3 mg | 6 IU | Ray et al. [4] |
| 10 | *Amaranthus viridis L.* | — | — | — | — | — | — | 330 mg | 18.7 mg | — | — | — | — | — | — | — | 179 mg | — | — | — | Patil et al. [35] |
| 11 | *Antidesma acidum Retz.* | 77.90% | — | 14.50 g | 6.3 g | 1.90 g | 15.76 g | 0.40 mg | 0.055 mg | 0.024 mg | — | 2.51 mg | 0.36 mg | 0.008 mg | 0.004 mg | 0.071 mg | — | — | — | — | Ray et al. [4] |
| 12 | *Aralia leschenaultii* (DC.) J.Wen | — | — | — | — | — | — | 0.31 mg | 0.6 mg | 0.25 mg | 0.47 mg | 1.89 mg | 0.6 mg | — | — | — | — | — | — | — | Chandran et al. [36] |
| 13 | *Aridisia macrocarpa* Wall | 67% | 154.20 kcal | 31.25 g | 6.6 g | — | 7.30 g | 76,952 mg | 100 mg | 2197 mg | 0.13 mg | 12,637 mg | 4347 mg | 5.7 mg | — | — | 4.44 mg | — | — | — | Ray et al. [4] |
| 14 | *Arisaema utile Hook.f. ex Schott* | — | — | — | — | — | — | 0.92 mg | 0.83 mg | 0.62 mg | 0.69 mg | 2.4 mg | 0.09 mg | — | — | — | — | — | — | — | Ray et al. [4] |
| 15 | *Artocarpus lakoocha Roxb. ex Buch.-Ham.* | — | — | — | — | — | — | 50 mg | 0.05 mg | — | — | — | — | — | — | — | 135 mg | — | 0.15 mg | — | Pandey Y et al. [37] |

| S. No. | Name of Species | Moisture | Energy | Protein | Fiber | Fat | Carbohydrate | Ca | Fe | Mg | P | K | Na | Zn | Cu | Mn | Vit C | B1 | B2 | Vit E | References |
|---|---|---|---|---|---|---|---|---|---|---|---|---|---|---|---|---|---|---|---|---|---|
| 16 | *Baccaurea sapida Lour.* | 81.17% | 377.44 kcal | 5.43 g | 3.6 g | 1.24 g | 86.14 g | 0.02377 mg | 0.02955 mg | 0.02167 mg | 0.13 mg | 0.37537 mg | 0.00799 mg | 0.00097 mg | 0.01433 mg | 0.03729 mg | 0.27 mg | — | — | — | Chandramouli et al. [38] |
| 17 | *Bambusa bambos (L.) Voss* | 92.20% | | 3.45 g | 3.7 g | 0.00020 g | 6.83 g | 5.43 mg | 2.23 mg | — | — | 367.33 mg | 12.23 mg | 0.82 mg | 0.36 mg | 0.94 mg | 5 mg | 0.1 mg | 0.19 mg | — | Ray et al. [4] |
| 18 | *Bambusa tulda Roxb.* | — | — | — | — | — | — | 4.06 mg | 3.19 mg | — | — | 408 mg | 19.96 mg | 0.72 mg | — | — | 1 mg | — | — | — | Ray et al. [4] |
| 19 | *Bauhinia purpurea L.* | — | — | — | — | — | — | 312 mg | — | — | — | — | — | — | — | — | — | — | — | — | Beegum et al. [39] |
| 20 | *Boerhaavia diffusa L.* | 78.90% | | 0.00576 g | 2.40 g | 0.00161 g | 0.01056 g | 202 mg | 10.68 mg | — | — | — | 39.4 mg | 0.41 mg | — | — | 0.20 mg | 0.24 mg | — | 16 IU | Ray et al. [4] |
| 21 | *Cannabis sativa L.* | — | — | — | — | — | — | 145 mg | 14 mg | 483 mg | — | 859 mg | 12 mg | 7 mg | — | — | — | — | 0.1 mg | — | Ray et al. [4] |
| 22 | *Carissa carandas L.* | — | — | — | — | — | — | 21 mg | — | — | — | — | — | — | — | — | — | — | — | — | Ray et al. [4] |
| 23 | *Castanopsis indica A.DC.* | — | — | — | — | — | — | 1540 mg | 2.6 mg | 12.68 mg | 80 mg | 4.33 mg | 0.03 mg | 1.53 mg | — | — | — | — | — | — | Fayaz et al. [40] |
| 24 | *Celosia argentea L.* | 81.20% | — | 0.0714 g | — | — | 0.0162 g | 86.6 mg | 0.522 mg | 67.18 mg | — | 37.66 mg | 0.99 mg | 0.0609 mg | — | 0.643 mg | — | — | — | — | Ray et al. [4] |
| 25 | *Centella asiatica (L.) Urb.* | — | — | — | — | — | — | 231 mg | 55.66 mg | — | — | — | 5.2 mg | 1.92 mg | — | — | — | 0.5 mg | — | 1 IU | Ray et al. [4] |
| 26 | *Chenopodium album L.* | — | — | — | — | — | — | 150 mg | 4.2 mg | — | — | — | — | — | — | — | 35 mg | — | 0.14 mg | 2 IU | Dangal et al. [41] |
| 27 | *Choerospondias axillaris L.* | — | 284 kcal | 3.87 g | 10.95 g | 0.88 g | 83.21 g | 1.58 mg | 0.11 mg | 0.68 mg | 0.16 mg | 0.67 mg | 0.04 mg | 0.83 mg | — | — | 67 mg | — | — | — | Ray et al. [4] |
| 28 | *Colocasia antiquorum (L.) Schott* | — | — | — | — | — | — | 227 mg | 10 mg | — | — | — | — | — | — | — | 12 mg | 0.2 mg | 0.26 mg | 6 IU | Temesgen et al. [42] |
| 29 | *Colocacia esculenta (L.) Schott* | 68.10% | — | 0.34 g | 2.50 g | 0.11 g | 26.80 g | 19 mg | 1.1 mg | 28 mg | — | 340 mg | 1 mg | 1.7 mg | — | — | 14.3 mg | 0.028 mg | 0.029 mg | — | Ray et al. [4] |
| 30 | *Commelina benghalensis L.* | — | — | — | — | — | — | 1431 mg | 115 mg | 7.98 mg | 220 mg | 390 mg | 200 mg | 2.68 mg | — | — | — | — | — | — | Katiyar et al. [43] |

| S. No. | Name of Species | Moisture | Energy | Protein | Fiber | Fat | Carbohydrate | Ca | Fe | Mg | P | K | Na | Zn | Cu | Mn | Vit C | B1 | B2 | Vit E | References |
|---|---|---|---|---|---|---|---|---|---|---|---|---|---|---|---|---|---|---|---|---|---|
| 31 | *Crotalaria juncea L.* | — | | 19.15 g | 30.22 g | | | 200 mg | — | — | — | — | — | — | — | — | | — | | — | Nayak et al. [44] |
| 32 | *Dendrocalamus strictus (Roxb.) Nees* | 0.50% | 296 kcal | 26.28 g | 17.42 g | 4.83 g | 36.84 g | 139.5 mg | 2.91 mg | 0.17 mg | 58.13 mg | — | 0.08 mg | — | — | — | 2 mg | — | — | — | Ray et al. [4] |
| 33 | *Dioscorea bulbifera L.* | — | — | — | — | — | — | 20 mg | 4.09 mg | — | — | — | 0.08 mg | 0.38 mg | — | — | 4 mg | — | — | — | Ray et al. [4] |
| 34 | *Diospyros melanoxylon Roxb.* | — | — | — | — | — | — | 60 mg | 0.5 mg | — | — | — | — | — | — | — | 1 mg | — | 0.04 mg | — | Ray et al. [4] |
| 35 | *Diplazium esculentum (Retz.) Sw.* | — | — | — | — | — | — | 1.02 mg | 0.56 mg | 0.51 mg | 0.5 mg | 2.37 mg | 0.08 mg | 0.58 mg | — | — | — | — | — | — | Ray et al. [4] |
| 36 | *Diploknema butyraceae a (Roxb.) H. J. Lam (Chyura)* | — | — | — | — | — | — | 0.82 mg | 0.18 mg | 0.61 mg | 0.09 mg | 0.82 mg | 0.07 mg | 0.86 mg | — | — | — | — | — | — | Ray et al. [4] |
| 37 | Elaeagnus latifolia L. | — | — | — | — | — | — | 1.47 mg | 0.18 mg | 0.54 mg | 0.09 mg | 0.91 mg | 0.05 mg | 1.19 mg | — | — | — | — | — | — | Ray et al. [4] |
| 38 | *Elaeagnus rhamnoides L.* | — | — | — | — | — | — | 0.17 mg | 0.06 mg | 0.31 mg | 0.31 mg | — | — | 0.88 mg | — | — | — | — | — | — | Pandey Y et al. [45] |
| 39 | *Elaeocarpus sikkimensis* Mast. | 68.67% | 389.56 kcal | 6.93 g | 2 g | 1.02 g | 88.17 g | 0.00289 mg | 0.02429 mg | 0.02167 mg | 0.07 mg | 0.22359 mg | 0.00809 mg | 0.00033 mg | 0.00427 mg | 0.01146 mg | — | — | — | — | Ray et al. [4] |
| 40 | *Ensete superbum (Roxb.) Cheesman* | — | — | — | — | — | — | 665.6 mg | 518 mg | 176 mg | — | 180 mg | 600 mg | 3.78 mg | — | — | — | — | — | — | Ray et al. [4] |
| 41 | *Enydra fluctuans* Lour. | — | — | — | — | — | — | 246 mg | 16.99 mg | — | — | — | 80 mg | 0.94 mg | — | — | 4 mg | 1 mg | — | 1 IU | Ray et al. [4] |
| 42 | *Euphorbia granulata* Forssk. | — | — | — | — | — | — | 425 mg | 81.09 mg | — | — | — | 24.9 mg | 1.01 mg | — | — | 9 mg | 3.1 mg | — | 12 IU | Govindan et al. [46] |
| 43 | *Ficus benghalensis L.* | — | 134.92Cal | 15.02 g | 19.45 g | — | 12.95 g | 364 mg | 33.46 mg | 220.7 mg | 250.9 mg | 192.3 mg | 399.4 mg | 11.34 mg | 12.19 mg | 21.55 mg | 203 mg | — | — | — | Saklani et al. [47] |
| 44 | *Ficus geniculata L.* | 46.64% | 135.51 kcal | 5.32 g | 16.96 g | 0.65 g | 27.09 g | 1.35 mg | 8.89 mg | 0.9 mg | 0.28 mg | 2.11 mg | 11.3 mg | 4.63 mg | — | — | 0.09 mg | — | — | 1 IU | Bhogaonkar et al. [48] |
| 45 | *Ficus racemosa L.* | 80.20% | — | 28.12 g | 0.544 g | 1.079 g | 15.84 g | 30.5 mg | 315 mg | 19.62 mg | 1312 mg | 49.3 mg | 329 mg | 0.05 mg | — | — | 0.0053 g | — | — | — | Ray et al. [4] |

| S. No. | Name of Species | Moisture | Energy | Protein | Fiber | Fat | Carbohydrate | Ca | Fe | Mg | P | K | Na | Zn | Cu | Mn | Vit C | B1 | B2 | Vit E | References |
|---|---|---|---|---|---|---|---|---|---|---|---|---|---|---|---|---|---|---|---|---|---|
| 46 | *Ficus sp.* | — | — | — | — | — | — | 295 mg | 2.77 mg | — | — | — | 7.5 mg | 0.8 mg | — | — | — | — | — | 8 IU | Singh and Khan [49] |
| 47 | *Hibiscus sabdariffa L.* | — | — | 3.5 g | — | 0.3 g | 8.7 g | 240 mg | 5 mg | — | — | — | — | 0.27 mg | — | — | 2.3 mg | 0.2 mg | 0.4 mg | 7 IU | Umar et al. [50] |
| 48 | *Ipomea aquatica Forssk* | 72.83% | 300.94 kcal | 6.30 g | 17.67 g | | 54.20 g | 416.70 mg | 210.30 mg | 301.64 mg | 109.29 mg | 5458.99 mg | 135 mg | 2.47 mg | 0.36 mg | 2.14 mg | 37 mg | 0.1 mg | 0.13 mg | 2 IU | Srivastava et al. [51] |
| 49 | *Kaempferia galanga L.* | 11.08% | 331.49 kcal | 5.92 g | 7.93 g | 1.66 g | 72.04 g | 950 mg | 69.91 mg | 293.92 mg | 60 mg | 12.23 mg | 0.32 mg | 8.35 mg | — | — | — | — | — | — | Ray et al. [4] |
| 50 | *Leucas cephalotes (Roth) Spreng.* | — | — | — | — | — | — | 236 mg | 20.02 mg | — | — | — | 10.6 mg | 0.8 mg | — | — | 8 mg | — | — | 18 IU | Ray et al. [4] |
| 51 | *Madhuca neriifoliaH.J.Lam* | — | — | — | — | — | — | 45 mg | 0.23 mg | — | — | — | — | — | — | — | 40 mg | — | — | — | Ray et al. [4] |
| 52 | *Marsilea minuta L.* | — | — | — | — | — | — | 53 mg | — | — | — | — | — | — | — | — | — | — | — | — | Bhamare et al. [52] |
| 53 | *Meyna spinose Roxb.* | 7.48% | — | 2.40300 g | — | — | — | 76.95 mg | 17.5 mg | 228.6 mg | 183.2 mg | 784.13 mg | 11.84 mg | 2.57 mg | 2.43 mg | 0.77 mg | 59.53 mg | — | — | — | Abbas et al. [53] |
| 54 | *Moringa oleifera Lam.* | — | — | 8.1 g | 2.1 g | 1.7 g | 9.1 g | 99.1 mg | 1.3 mg | 35.1 mg | 70.8 mg | 471 mg | 70 mg | 0.85 mg | — | — | 8.6 mg | 0.103 mg | 0.112 mg | 266.4 IU | Ray et al. [4] |
| 55 | *Nelumbo nucifera Gaertn.* | 48.85% | 40 kcal | 0.95 g | 1.9 g | 0.04 g | 9.61 g | 16 mg | 0.54 mg | 13 mg | 47 mg | 218 mg | 27 mg | 0.2 mg | 0.13 mg | 0.132 mg | 16.4 mg | 0.076 mg | 0.006 mg | — | Anand et al. [54] |
| 56 | *Nymphaea nouchali Burm.f.* | 9.07% | — | 10.76 g | 0.64 g | 2.40 g | 76.50 g | 148.55 mg | 1.98 mg | 145.48 mg | — | 858.39 mg | 643.58 mg | 1.33 mg | — | — | 3.12 mg | 0.05 mg | 1.11 mg | — | Stephen et al. [55] |
| 57 | *Nympheae rubra Burm.f.* | 9.72% | — | 21.66 g | 13.30 g | 5.07 g | 41.92 g | 354.1 mg | 41.38 mg | 87.46 mg | 635.39 mg | 742.89 mg | 431.53 mg | 8.16 mg | 12.36 mg | — | 24.650 mg | 0.085 mg | 0.065 mg | 1.71028.8 IU | Jain et al. [56] |
| 58 | *Oxalis corniculata L.* | 82.42% | — | 22.28 g | — | — | 24.67 g | 2500 mg | 234 mg | 250 mg | — | 2170 mg | 1120 mg | 14.75 mg | — | — | 21 mg | — | — | — | Hasan et al. [57] |
| 59 | *Phyllanthus embilica L.* | 81.20% | — | 0.50 g | 3.40 g | 0.10 g | 14.10 g | 27.6 mg | 3.3 mg | 11.8 mg | 28.2 mg | 282 mg | 4.2 mg | 1.8 mg | — | — | 600 mg | — | | — | Ray et al. [4] |
| 60 | *Polygonum molle D. Don* | — | — | — | — | — | — | 27.6 mg | 3.3 mg | 11.8 mg | 28.2 mg | 282 mg | 4.2 mg | 1.8 mg | — | — | — | — | — | — | Ray et al. [4] |

| S. No. | Name of Species | Moisture | Energy | Protein | Fiber | Fat | Carbohydrate | Ca | Fe | Mg | P | K | Na | Zn | Cu | Mn | Vit C | B1 | B2 | Vit E | References |
|---|---|---|---|---|---|---|---|---|---|---|---|---|---|---|---|---|---|---|---|---|---|
| 61 | *Polygonum plebeium* R.Br. | — | — | — | — | — | — | 0.15 mg | 0.32 mg | 0.43 mg | 0.27 mg | 2.02 mg | 0.09 mg | 0.28 mg | — | — | | — | | — | Uddin et al. [58] |
| 62 | *Portulaca olearaceae* L. | — | 16 kcal | 1.30 g | — | 0.1 g | 3.4 g | 65 mg | 1.99 mg | 68 mg | 44 mg | 494 mg | 45 mg | 0.17 mg | 0.113 mg | 0.303 mg | 21 mg | 0.047 mg | 0.112 mg | 1320 IU | Ray et al. [4] |
| 63 | *Prunus cerasoides* Buch.-Ham. ex D. Don | — | — | — | — | — | — | 0.2 mg | 0.21 mg | 0.59 mg | 0.18 mg | 0.47 mg | 0.02 mg | 0.2 mg | — | — | — | — | — | — | Ray et al. [4] |
| 64 | *Prunus napaulensis* (Ser.) Steud. | — | — | — | — | — | — | 1220 mg | 10.7 mg | 217.74 mg | 70 mg | 16.5 mg | 0.1 mg | 1.49 mg | — | — | 608 mg | — | — | — | Ray et al. [4] |
| 65 | *Quercus robur* L. | 7.91% | 110 kcal | 1.74 g | | 6.76 g | 11.55 g | 12 mg | 0.2 mg | 18 mg | 22 mg | 153 mg | 0.13 mg | 0.14 mg | 0.176 mg | 0.379 mg | — | 0.032 mg | 0.033 mg | 11 IU | Ray et al. [4] |
| 66 | *Rhus chinensis* Mill. | — | — | — | — | — | — | 1020 mg | 4.16 mg | 111 mg | 160 mg | 8.41 mg | 0.03 mg | 2.37 mg | — | — | — | — | — | — | Ray et al. [4] |
| 67 | *Schleichera trijuga* (Lour.) Oken | — | — | — | — | — | — | 15 mg | — | — | — | — | — | — | — | — | — | — | — | — | Ray et al. [4] |
| 68 | *Semecarpus anacardium* L.f. | — | — | — | — | — | — | 295 mg | 6.1 mg | — | — | — | — | — | — | — | — | — | — | — | Ray et al. [4] |
| 69 | *Senna tora* (L.) Roxb. | — | — | — | — | — | — | 520 mg | 12.4 mg | — | — | — | — | — | — | — | 82 mg | 0.1 mg | 0.19 mg | 11 IU | Abraham et al. [59] |
| 70 | *Solanum torvum* Sw. | 79.81% | — | 10.72 g | 25.12 g | 8.81 g | 0.33 g | 0.47 mg | 208.43 mg | 0.32 mg | 0.58 mg | 1.84 mg | 0.31 mg | 16.58 mg | 14.97 mg | 102.83 mg | 3 mg | — | — | — | Ray et al. [4] |
| 71 | *Spondia pinnata* (L. f.) Kurz | — | — | — | — | — | — | 0.93 mg | 1.32 mg | — | 0.68 mg | 1.38 mg | 1.54 mg | — | — | — | — | — | — | — | Kuru et al. [60] |
| 72 | *Tamarindus indica* L. | — | 239 kcal | 2.80 g | 5.10 g | 0.60 g | 62.5 g | 74 mg | 2.8 mg | 92 mg | 113 mg | 628 mg | 28 mg | 0.10 mg | 0.86 mg | — | 3.5 mg | 0.428 mg | — | 30 IU | Ray et al. [4] |
| 73 | *Terminalia chebula*L . | — | — | — | — | — | — | 0.81 mg | 0.03 mg | 0.3 mg | 0.04 mg | 1.27 mg | 0.08 mg | 0.44 mg | — | — | — | — | — | — | Khan et al. [61] |
| 74 | *Trianthema portulacastrum* L. | 0.08% | 76.01 kcal | 0.0919 g | 0.43 g | | 0.0302 g | 100 mg | 6.44 mg | — | — | 51.60 mg | 44 mg | 0.20 mg | 0.02 mg | 0.04 mg | — | — | 2.02 mg | 2697.3 IU | Ray et al. [4] |
| 75 | *Urtica dioica* L. | — | — | — | — | — | — | 1.31 mg | 1.31 mg | 0.27 mg | — | 1.87 mg | 0.07 mg | — | — | — | — | — | — | — | Ray et al. [4] |

DOI: http://dx.doi.org/10.5772/intechopen.1009820

| S. No. | Name of Species | Moisture | Energy | Protein | Fiber | Fat | Carbohydrate | Ca | Fe | Mg | P | K | Na | Zn | Cu | Mn | Vit C | B1 | B2 | Vit E | References |
|---|---|---|---|---|---|---|---|---|---|---|---|---|---|---|---|---|---|---|---|---|---|
| 76 | *Viburnum corylifolium Hook.f. & Thomson* | — | — | — | — | — | — | 630 mg | 3.55 mg | 161.39 mg | 140 mg | 11.13 mg | 0.11 mg | 1.62 mg | — | — | 238 mg | — | — | — | Ray et al. [4] |
| 77 | *Vicia hirsute (L.) Gray* | — | — | — | — | — | — | 215 mg | 7.78 mg | — | — | — | 33.18 mg | 4.11 mg | — | — | 23 mg | — | — | 1 IU | Ray et al. [4] |
| 78 | *Zanthoxylum rhetsa (Roxb.) DC.* | — | — | — | — | — | — | 0.88 mg | 0.05 mg | 0.35 mg | 0.14 mg | 0.72 mg | 0.02 mg | 1.16 mg | — | — | — | — | — | — | Pareek et al. [62] |
| 79 | *Ziziphus jujuba Mill.* | 83.0% | — | 0.8 g | 0.60 g | 0.07 g | 17 g | 25.6 mg | 1.8 mg | — | 26.8 mg | — | — | 0.1 mg | — | — | 76 mg | 0.024 mg | 0.038 mg | — | Ray et al. [4] |

**Table 1.**
*Nutritional composition of selected wild edible plants, illustrating their contributions in terms of proteins, fats, carbohydrates, and minerals.*

*Elatostema platyphyllum, Gynura bicolor, Plantago erosa, and Diplazium esculentum* were widely consumed as vegetables by the ethnic tribes of Arunachal Pradesh. Among them *Diplazium esculentum* was richest in fiber and carbohydrate content fetching the greatest energy value with rich minerals, whereas ash content was highest in *Elatostema platyphyllum* and lowest in *Polygonum runcinatum*. Crude protein ranged from 1.74–2.50 g/100 g edible portions, and it was highest in *P. runcinatum* and lowest in *Gynura bicolor*, whereas fat content ranged from 0.12–0.20 g/100 g edible portions with the highest content in *G. bicolor* and lowest in *D. esculentum*. The energetic value of the six plants ranged from 31.0–37.7 kcal/100 g edible portions, which was highest in *D. esculentum* and lowest in *G. bicolor* [64]. In Meghalaya, the bulb of *Dioscorea bulbifera*, the underground stem of *Homalomena aromatica*, the flower of *Phlogacanthus curviflorus*, the leaf of *Medinilla erythrophylla* and *Ardisia humilis*, and fruits of *Careya arborea* were used by the tribal people. In a study, the crude fat content was reported with an average of 0.17–1.39%, and crude protein content was found to be the highest in the leaves of *A. humilis* (12.71%) while the available carbohydrate content was the highest in the fruits of *Careya arborea* (88.08%). Amaranthus species are a good source of potassium, iron, zinc, magnesium, and particularly, calcium with appreciable amounts of carotenes and vitamin C, and they were rated above cabbage and spinach in terms of nutritional characteristics. This nutritional information can be used by health departments, public agencies, and agricultural industries to promote the commercialization of wild edible plants to sustain global nutritional security.

## 5. Wild edible plants safeguard food security and nutritional adequacy

Food and nutrition security is a crucial issue that our nation is now struggling with. Micronutrient deficiencies are estimated to affect billions of individuals. Food security is a serious issue in developing countries, where high cost, scarcity, and erratic supply of nutritious food have prompted a search for low-cost and alternative sources of nutritional meals. Edible wild plants are one of the additional sources of healthy meals and ensure both affordable food and nutritional security. These edible wild plants have played an essential part in safeguarding food security and nutritive variation. They have remarkable nutritional content and can be a major source of proteins and vitamins A, B, and C. These are also powerful substitutes for traditional plant-based diets and have therapeutic benefits. Wild plants have the potential to boost food security by providing a wide variety of affordable and nutritious food sources that are available throughout the year and can develop in diverse environmental conditions. Edible wild plants have the potential to meet dietary demands and provide balanced nutrition if consumed in recommended portions and sizes. Food insecurity and malnutrition impact nearly one in every seven people globally, while growth in population, rising consumption, and climate change all threaten to raise the risks of hunger in the future. Wild plants are an important source of food in India, and they play a significant role in the nutrition of small and marginal farmer households, as well as forest dwellers, during times of food scarcity. Tribal eating habits are mostly determined by the seasonal availability of food and its nutritional value. Traditional knowledge of edible wild plants has been passed down through the generations through word of mouth. The younger generation learns to recognize the plants and plant parts gathered while accompanying their parents to the forest [65].

The best method of improving household food security focuses on natural resources. Edible wild plants are well-known to be used by over one billion people as a means to supplement protein and vitamins, enhance the palatability of staple meals, and generate revenue. During times of severe food scarcity, consumers of wild edibles can diversify their food sources, mitigate malnutrition, and generate additional revenue. In India, wild fruits and veggies play a major role in the food and nutritional security of rural and indigenous peoples. Some wild plants are known to be nutritionally superior to produced fruits. Considering the significance of edible wild plants to household food security, the social–ecological systems that allow for the collection of these natural resources must be properly preserved, managed, and valued to avoid overexploitation and deterioration [7, 10].

## 6. Medicinal importance of wild edible fruits and vegetables

WEPs are used for medicinal purposes to treat several diseases and is marketed to provide revenue for rural communities; there are around 40 vitamins and minerals that are thought to be necessary for physical and mental growth, as well as the immune system and metabolic activities. In diverse human civilizations across the world, more than 20,000 plant species are consumed for medicinal purposes [66]. Due to their high antioxidant and fiber content, wild fruits and vegetables have the curing ability to treat several disorders, including diabetes, cardiovascular disease, inflammation, and digestive and urinary tract diseases. WEPs are edible and contain nutritional value, as they contain minerals such as sodium, magnesium, potassium, calcium, iron, and phosphorus [67]. They are effective against various diseases and are often used in different Ayurveda formulations in Indian Folk-medicine.

Due to the overuse of artificial fertilizers, many conventional vegetables and fruits are losing their natural taste and nutritional content [68]. On the other hand, due to the poor production of foods, some people are suffering from hunger, and therefore, many people are consuming ethnic food plants to fulfill their nutritional requirements. These ethnic food plants not only provide food but also have a variety of medical properties. Since ancient times, indigenous tribal peoples have consumed wild plants across the planet [69]. In the developing world, wild plants serve as a significant food source, medicine, and material subsistence. In the past few years, there has been a growth in the desire to know more about the traditions of utilizing wild plants beyond material purposes and focusing on local nutrition, medicinal purposes, dietary diversity, income generation, health care, micronutrient deficiency reduction, and food security through diversification.

WEPs are a key source of food and income for the rural population, and they are also therapeutic in nature and are used to cure a range of illnesses. Therefore, special consideration should be given to preserving and improving this vital source of food, as well as conserving it in the future.

## 7. Securing the future for wild foods

The first step in promoting good eating habits is to raise awareness of food nutritive value, proper nutrition, and healthy eating habits. Wild edible plants are the precious gift of nature. They played a significant role in serving the nutritional needs of local people in remote areas of the country. Residents consume a variety of WEPs in

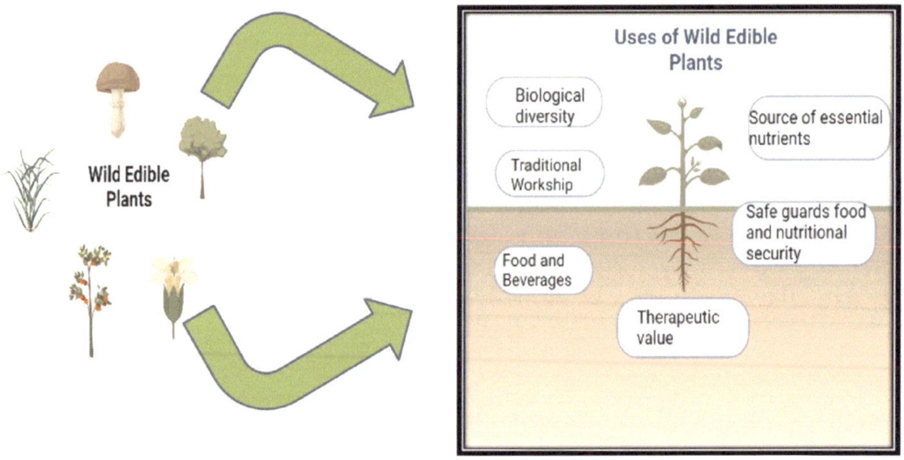


**Figure 1.**
*Graphical representation of the dimensions and properties of the wild edible plants.*

various forms such as tubers, roots, leaves, flowers, fruits, and seeds [9]. Wild food plants are more commonly used by poor and marginalized rural families and farming communities. These plants can give extra revenue to rural households through collection and sale in local marketplaces [70]. Many wild edible plants are more nutritious than many farmed varieties [7]. Unfortunately, due to the arrival of industrialized and processed food items, the utilization of indigenous food plants in rural regions has fallen out of favor. There are several threats and barriers to the utilization of edible wild plants, viz. land-use changes of natural lands to agricultural land, urbanization activities, climate change hazards, unscientific harvesting, and uncontrolled grazing.

Increased intake of processed food can be harmful to human health. Malnutrition (encompassing both excess and undernutrition) is seen as a worldwide concern, indicating a pressing need for a healthy and sustainable food system. WEPs can therefore serve as a crucial serving as a crucial component of people's diets in various parts of the globe, increasing nutritional diversity for those who rely on them. Food plants are sometimes consumed for their medicinal effects, and several species are frequently employed as herbal medicines in folk treatments to diagnose a range of ailments [71]. They are frequently characterized as functional foods due to their greater amount of vitamins, phenols, flavonoids, antioxidants, microelements, and fiber than farmed crops and their evident favorable impact on health. Wild plants are also seen as a healthier option than cultivated crops, which may contain pesticides and other pollutants. As a result, wild species may offer a lot of promises as sources of distinctive color and taste, as well as bioactive chemicals and nutritional supplements. Edible wild plants provide a substantial contribution to food security, nutrition, and livelihoods. The function of WFPs might be reinforced in the future with more development focus and research funding, as well as a more efficient enabling policy environment. Micronutrients are abundant in many of these plants, which can be rare in high-global-production crops. In addition to the health and dietary benefits, wild food plants are pesticide and fertilizer-free. In comparison with conventionally grown plants, wild edible fruits and vegetables growing in their natural conditions need less maintenance and are rich sources of dietary fibers, micronutrients, vitamins, and minerals that are essential for good health and boost immunity against infectious diseases. There is a need to diversify the diet including traditionally acceptable wild

*DOI: http://dx.doi.org/10.5772/intechopen.1009820*

edible plants through scientific interventions. Hence, different healthcare programs regarding what to eat to keep healthy must be developed, with a focus on including the lesser-known edible wild fruits and vegetables, which can be important nutrient contributors to the diet and help to prevent nutritional deficiency and enhance food security (**Figure 1**).

## 8. Livelihood opportunities from WEPs

Wild edible plant diets vary from region to region. Within one country itself, many varieties can be found. The dietary intake of wild edible plants may vary demographically, and some are consumed only during traditional festival occasions. Upon commercialization, wild edible plants will get worldwide recognition. This will be beneficial for the human race nutritionally. Many wild plant-based foods are becoming scarce and very much endemic to any particular place. Tourism, especially eco-tourism, can have a major stake in introducing various edible local floras to the tourists, and local people can earn an income out of it [72]. Such endeavors can also help in giving international recognition to lesser-known wild edible food plants. Globally, changes in food consumption patterns have caused a reduction in food dependency on once common wild edible plants. The shift in consumption paradigm is also affected by the economic status of the families as many of these are considered famine foods. But in reality, these foods are a rich source of macronutrients, and they can alleviate hidden hunger. The deficiency of micronutrients is mostly unrecognized and ignored in communities though one-third of the world's human population is affected. Several studies have reported the importance of wild edible foods in enhancing food security. Wild foods play a great role in supporting rural households though they are not getting enough recognition in policymaking. United Nations SDGs can make significant contributions to developing policy interventions in WEPs. Many non-quantifications of the economic contribution of dietary wild plants have been reported by [12] suggesting that wild foods have the potential to increase household food security upon integration. Generally, WEPs are produced in homesteads for subsistence purposes only. Value addition of wild edible food plants and their commercialization can benefit income generation and thereby livelihood upliftment in rural households. This will reduce the dependency on forests sustainably and increase the consumer base globally as people are slowly diverting toward eco-friendly or nature-based consumption patterns. Thereby we can help millions of people away from financial and micronutrient poverty.

## 9. Recommendations and future research prospects

Recommendations:

- Promote awareness and education: Develop programs to teach the identification and use of wild edible plants through schools and community outreach.

- Document traditional knowledge: Preserve indigenous knowledge on wild edible plants and share it with the public.

- Validate nutritional benefits: Conduct scientific research on the nutritional value of wild edible plants and integrate findings into dietary guidelines.

- Community-based cultivation: Encourage the creation of local gardens focusing on wild edible plants to improve food accessibility.
- Government nutrition programs: Incorporate wild edible plants into nutrition programs, particularly in rural and tribal regions.
- Commercialization: Foster public–private partnerships to create sustainable markets for wild edible plants.
- Conservation: Protect natural habitats to ensure the survival of wild edible plants.
- Healthcare training: Educate healthcare professionals on using wild edible plants to address nutritional deficiencies.
- Digital platforms: Utilize mobile apps and digital tools to spread knowledge on wild edible plants.
- Culinary innovations: Create new recipes and food products to boost the popularity of wild edible plants.

Future research prospects:

- Nutritional profiles: Explore the nutrient composition and bioavailability in underutilized wild edible plants.
- Ethnobotanical studies: Investigate traditional uses of wild edible plants for medicinal purposes and their role in reducing micronutrient deficiencies.
- Sustainable practices: Study sustainable harvesting and cultivation methods to preserve wild edible species.
- Health outcomes: Conduct long-term studies to understand the health benefits of wild edible plants, particularly for non-communicable diseases.
- Food system resilience: Evaluate the potential of wild edible plants to strengthen food systems, particularly in the face of climate change.
- Safety and toxicology: Research the safety of consuming wild edible plants and their potential as functional foods.
- Ecosystem services: Assess the ecological benefits of wild edible plants and integrate them into national policies.
- Biotechnological enhancement: Use biotechnology to enhance the nutritional properties and cultivation of wild edible plants.

By addressing these recommendations and research prospects, India can effectively leverage the potential of wild edible plants to alleviate nutritional deficiencies and enhance food security, contributing to overall health and well-being of its population.

## 10. Conclusions

Wild edible plants (WEPs) improve food security by providing low-cost, nutritious food sources high in vitamins, minerals, fiber, and bioactive compounds. They thrive in a variety of environments, including water-stressed areas, making them critical during periods of food scarcity or limited market access. WEPs serve a variety of industries: pharmaceuticals use their medicinal compounds, cosmetics use their natural extracts, functional food companies create nutrient-dense products, textiles use their natural dyes, and some species show promise for biofuel production. These plants offer benefits beyond nutrition: they contribute to climate resilience, provide ecosystem services, create economic opportunities, address micronutrient deficiencies, and preserve cultural heritage. However, there are challenges: overharvesting endangers wild populations; toxicity and misidentification pose safety risks; habitat loss endangers many species; regulatory frameworks may impede commercialization; and knowledge gaps regarding many species persist. To maximize WEPs' long-term use, stakeholders should develop conservation strategies, establish equitable benefit-sharing mechanisms, create value chains that ensure fair returns, incorporate WEPs into national policies, invest in interdisciplinary research, develop certification systems, and create educational programs that combine traditional knowledge with scientific understanding.

## Author details

Aakanksha Pandey[1], Archana Mishra[2], Smaranika Mohanta[2], Abha Manohar Kandileri[2]*, Gopal Shukla[3] and Sumit Chakravarty[4]

1 Department of Extension Education and Communication Management, Punjab Agricultural University, Ludhiana, Punjab, India

2 M. S. Swaminathan School of Agriculture, Centurion University of Technology and Management, Odisha, India

3 Department of Forestry, North-Eastern Hill University, Tura Campus, Tura, Meghalaya, India

4 Department of Forestry, Uttar Banga Krishi Viswavidyalaya, Pundibari, West Bengal, India

*Address all correspondence to: abha.manohar@cutm.ac.in

## References


[1] Flyman MV, Afolayan AJ. The suitability of wild vegetables for alleviating human dietary deficiencies. South African Journal of Botany. 2006; 72:492-497

[2] Childs CE, Calder PC, Miles EA. Diet and immune function. Nutrients. 2019;**1** (8):1933

[3] Mahapatra AK, Panda PC. Wild Edible Fruit Plants of Eastern India. Bhubaneswar: Regional Plant Resource Centre; 2009. pp. 1-15

[4] Ray A, Ray R, Sreevidya EA. How many wild edible plants do we eat-their diversity, use, and implications for sustainable food system: An exploratory analysis in India. Frontiers in Sustainable Food Systems. 2020;**4**:56

[5] Singh TA, Sarangi PK, Sarangthem N. Food Bio Resources and Ethnic Foods of Manipur, North East India. Uttar Pradesh: Empyreal Publishing House; 2019. pp. 1-121

[6] Powell B, Thilsted S, Ickowitz A, Termote C, Sunderland T, Herforth A. Improving diets with wild and cultivated biodiversity from across the landscape. Food Security. 2015;7:535-554

[7] Duguma HT. Wild edible plant nutritional contribution and consumer perception in Ethiopia. International Journal of Food Science. 2020. Available from: https://www.hindawi.com/journals/ijfs/2020/2958623/

[8] Schippmann U, Cunningham AB, Leaman DJ. Impact of cultivation and gathering of medicinal plants on biodiversity: Global trends and issues. In: Biodiversity and the Ecosystem Approach in Agriculture, Forestry and Fisheries. Rome: FAO; 2002

[9] Uprety Y, Poudel RC, Shrestha KK, Rajbhandary S, Tiwari NN, Shrestha UB, et al. Diversity of use and local knowledge of wild edible plant resources in Nepal. Journal of Ethnobiology and Ethnomedicine. 2012;**8**:1-15

[10] Mishra A, Swamy SL, Thakur TK, Bhat R, Bijalwan A, Kumar A. Use of wild edible plants: Can they meet the dietary and nutritional needs of indigenous communities in Central India. Food. 2021;**10**:14-53

[11] Misra RC, Sahoo HK, Pani DR, Bhandari DC. Genetic resources of wild tuberous food plants traditionally used in Similipal biosphere reserve, Odisha, India. Genetic Resources and Crop Evolution. 2013;**60**:2033-2054

[12] Hickey GM, Pouliot M, Smith-Hall C, Wunder S, Nielsen MR. Quantifying the economic contribution of wild food harvests to rural livelihoods: A global-comparative analysis. Food Policy. 2016; **62**:122-132

[13] FAO. Forests, Trees and People Together in a Living Landscape: A Key Rural Development Committee on Forestry Paper COFO/2012/6.2. Rome, Italy; 2012. Available from: http://www.fao.org/docrep/meeting/026/me435e.pdf

[14] Aberoumand A, Deokule SS. Studies on nutritional values of some wild edible plants from Iran and India. Pakistan Journal of Nutrition. 2009;**8**:26-31

[15] Kumar A. Ethnobotanical study of wild vegetables used by rural communities of Kannauj district, Uttar Pradesh, India. Emirates Journal of Food and Agriculture. 2013;**24**:760-766

[16] Bacchetta L, Visioli F, Cappelli G, Caruso E, Martin G, Nemeth E, et al.

A manifesto for the valorization of wild edible plants. Journal of Ethnopharmacology. 2016;**191**: 180-187

[17] Ruane J, Sonnino A. Agricultural biotechnologies in developing countries and their possible contribution to food security. Journal of Biotechnology. 2011; **156**:356-363

[18] Singh B, Sultan P, Hassan QP, Gairola S, Bedi YS. Ethnobotany, traditional knowledge, and diversity of wild edible plants and fungi: A case study in the Bandipora district of Kashmir Himalaya, India. Journal of Herbs Spices & Medicinal Plants. 2016; **22**:247-278

[19] Borelli T, Hunter D, Powell B, Ulian T, Mattana E, Termote C, et al. Born to eat wild: An integrated conservation approach to secure wild food plants for food security and nutrition. Plants. 2020; **9**(10):12-99

[20] Karasawa MMG, Mohan C. Fruits as prospective reserves of bioactive compounds: A review. Natural Products and Bioprospecting. 2018;**8**:335-346

[21] Joshi SK, Ballabh B, Negi PS, Dwivedi SK. Diversity, distribution, use pattern and evaluation of wild edible plants of Uttarakhand, India. Defence Life Science Journal. 2018;**3**:126-135

[22] Oguntibeju OO, Truter EJ, Esterhuyse AJ. The role of fruit and vegetable consumption in human health and disease prevention. Diabetes Mellitus-Insights and Perspectives. 2013; **3**:172-180

[23] Dembitsky VM, Poovarodom S, Leontowicz H, Leontowicz M, Vearasilp S, Trakhtenberg S, et al. The multiple nutrition properties of some exotic fruits: Biological activity and active metabolites. Food Research International. 2011;**44**:1671-1701

[24] Khan A, Khan S, Khan MA, Qamar Z, Waqas M. The uptake and bioaccumulation of heavy metals by food plants, their effects on plants nutrients, and associated health risk: A review. Environmental Science and Pollution Research. 2015;**22**:13772-13799

[25] Saikia P, Chandra Deka D. Analysis of nutrient content of five non-conventional vegetables of Assam, India. Mediterranean Journal of Nutrition and Metabolism. 2015;**8**:101-108

[26] Satter MMA, Khan MMRL, Jabin SA, Abedin N, Islam MF, Shaha B. Nutritional quality and safety aspects of wild vegetables consume in Bangladesh. Asian Pacific Journal of Tropical Biomedicine. 2016;**6**:125-131

[27] Joshipura H. The effect of fruit and vegetable intake on risk for coronary heart disease. Annals of Internal Medicine. 2001;**134**:1106-1114

[28] Al-Farga A, Zhang H, Siddeeg A, Shamoon M, Chamba MV, Al-Hajj N. Proximate composition, functional properties, amino acid, mineral and vitamin contents of a novel food: Alhydwan (*Boerhavia elegana* Choisy) seed flour. Food Chemistry. 2016;**211**: 268-273

[29] Natesh HN, Abbey L, Asiedu SK. An overview of nutritional and anti-nutritional factors in green leafy vegetables. Horticulture International Journal. 2017;**1**:00-11

[30] Omoyeni OA, Adeyeye EI. Chemical composition, calcium, zinc and phytate interrelationships in *Aerva lanata* (Linn) Juss. Ex schult leaves. Oriental Journal of Chemistry. 2009;**25**:485-488

[31] Mishra ML, Sood S, Shukla UN. Phyto-nutritional and mineral composition of Indian horse chestnut (*Aesculus indica*) seeds. Journal of Pharmacognosy and Phytochemistry. 2018;7:2159-2162

[32] Muszynska B, Kala K, Rojowski J, Grzywacz A, Opoka W. Composition and biological properties of *Agaricus bisporus* fruiting bodies-a review. Polish Journal of Food And Nutrition Sciences. 2017;**67**(3):173-181

[33] Leung WTW, Busson F, Jardin C. Food Composition Table for Use in Africa. FAO; 1968

[34] Soriano-García M, Aguirre-Díaz IS. Nutritional functional value and therapeutic utilization of Amaranth. In: Nutritional Value of Amaranth. London, UK: IntechOpen; 2019

[35] Patil VA, Nitave SA. A review on Eucalyptus globulus: A divine medicinal herb. World Journal of Pharmacy and Pharmaceutical Sciences. 2014;**3**(6): 559-567

[36] Chandran PR, Manju S, Vysakhi MV, Shaji PK, Nair AG. Nutritional and antinutritional properties of the leaf of *Ardisia solanacea* Roxb.(Myrsinaceae), a fodder additive. International Food Research Journal. 2015;**22**:324

[37] Pandey Y, Upadhyay S, Bhatt SS, Sharma L, Manivannan S, Chanbisana C. Nutritional compositions of *Baccaurea sapida* and Eleaocarpus sikkimnesis of Sikkim Himalaya. International Journal of Current Microbiology and Applied Sciences. 2018;7:2101-2106

[38] Chandramouli S, Viswanath S. Nutritional composition of edible bamboo shoots of some commercially important bamboo species in peninsular India. International Journal of Life Sciences. 2015;**3**:275-287

[39] Beegum GR, Beevy SS, Sugunan VS. Nutritive and anti-nutritive properties of *Boerhavia diffusa* L. Journal of Pharmacognosy and Phytochemistry. 2014;**2**:147-151

[40] Fayaz MUFIDA, Bhat MH, Kumar AMIT, Jain AK. Phytochemical screening and nutritional analysis of some parts of *Celosia argentea* L. Chemical Science. 2019;**8**:12-19

[41] Yadav KC, Dangal A, Thapa S, Rayamajhi S, Chalise K, Shiwakoti LD, et al. Nutritional, phytochemicals, and sensory analysis of Lapsi (*Choerospondias axillaris*) fruit leather. International Journal of Food Properties. 2022;**25**: 960-975

[42] Temesgen M, Retta N. Nutritional potential, health and food security benefits of taro *Colocasia esculenta* (L.): A review. Food Science and Quality Management. 2015;**36**:23-30

[43] Katiyar SK, Agarwal R. Antioxidant and anti-inflammatory properties of green tea polyphenols. Current Medicinal Chemistry. 1969;**6**(3):241-249

[44] Nayak BP, Palta A. Development of bamboo (*Dendrocalamus strictus*) shoot powder and its nutritional value. International Journal of Food and Nutritional Sciences. 2022;**11**:15-20

[45] Pandey MM, Khatoon S, Rastogi S, Rawat AKS. Determination of flavonoids, polyphenols and antioxidant activity of Tephrosia purpurea: A seasonal study. Journal of Integrative Medicine. 2016;**14**(6):447-455

[46] Govindan V, Shoba FG. Determination of nutritive value and mineral elements of *Ficus benghalensis*

Linn. Seed. Asian Journal of Pharmaceutical and Clinical Research. 2015;**8**:156-159

[47] Saklani S, Chandra S. In vitro antimicrobial activity, nutritional profile and phytochemical screening of wild edible fruit of Garhwal Himalaya (*Ficus auriculata*). International Journal of Pharmaceutical Sciences Review and Research. 2012;**12**:61-64

[48] Bhogaonkar PY, Devarkar VD. Ethnobotanical studies on medicinal plants of Melghat forest in Amravati district of Maharashtra. International Journal of Herbal Medicine. 2014;**2**(2): 6-11

[49] Singh P, Khan M, Hailemariam H. Nutritional and health importance of *Hibiscus sabdariffa*: A review and indication for research needs. Journal of Nutritional Health & Food Engineering. 2017;**6**:212

[50] Umar KJ, Hassan LG, Dangoggo SM, Ladan MJ. Nutritional composition of water spinach (Ipomoea aquatica Forsk) leaves. Journal of Applied Sciences. 2007;**7**(6):803-809

[51] Srivastava N, Singh S, Gupta AC, Shanker K, Bawankule DU, Luqman S. Aromatic ginger (*Kaempferia galanga* L.) extracts with ameliorative and protective potential as a functional food, beyond its flavor and nutritional benefits. Toxicology Reports. 2019;**6**:521-528

[52] Bhamare M, Shinde V, Khalkar K, Khandbahale D, Gaikwad K. Nutritional profiling of wild edible fruits of Meyna laxiflora Robyns. International Journal of Botany Studies. 2022;**7**(2):368-373

[53] Abbas RK, Elsharbasy FS, Fadlelmula AA. Nutritional values of Moringa oleifera, total protein. Amino acid, vitamins, minerals, carbohydrates, total fat and crude fiber, under the semi-arid conditions of Sudan. Journal of Microbial and Biochemical Technology. 2018;**10**(2):56-58

[54] Anand U, Tudu CK, Nandy S, Sunita K, Tripathi V, Loake GJ, et al. Ethnodermatological use of medicinal plants in India: From Ayurvedic formulations to clinical perspectives—A review. Journal of Ethnopharmacology. 2022;**284**:114744

[55] Stephen EC, Adebisi AK, Chinedu I, Samuel AA. Chemical composition of water lily (Nymphaea lotus) bulbs. American Journal of Food Science and Nutrition. 2017;**4**:7-12

[56] Jain DL, Baheti AM, Jain SR, Khandelwal KR. Use of medicinal plants among tribes in Satpuda region of Dhule and Jalgaon districts of Maharashtra—An ethnobotanical survey. Indian Journal of Traditional Knowledge. 2010;**9**(1): 152-157

[57] Hasan M, Abdullah HM. Plant genetic resources and traditional knowledge: Emerging needs for conservation. In: Plant Genetic Resources and Traditional Knowledge for Food Security. 2016. pp. 978-981

[58] Uddin M, Juraimi AS, Hossain MS, Nahar M, Un A, Ali M, et al. Purslane weed (*Portulaca oleracea*): A prospective plant source of nutrition, omega-3 fatty acid, and antioxidant attributes. Scientific World Journal. 2014;**2014**:1-6. Available from: https://ndb.nal.usda.gov/ndb/foods/show/3

[59] Abraham JD, Sekyere EK, Gyamerah I. Effect of boiling on the nutrient composition of *Solanum Torvum*. International Journal of Food Science. 2022;**2022**(1):7539151. DOI: 10.1155/2022/7539151

[60] Kuru P, Kuru P. Medicinal plants used in traditional medicine in Turkey. Journal of Ethnopharmacology. 2014;**155** (1):1-8

[61] Khan N, Sultana A, Tahir N, Jamila N. Nutritional composition, vitamins, minerals and toxic heavy metals analysis of *Trianthema portulacastrum* L., a wild edible plant from Peshawar, Khyber Pakhtunkhwa, Pakistan. African Journal of Biotechnology. 2013;**12**(42):6079-6085

[62] Pareek S, Sharma R, Meena R. Ethnobotanical study of wild edible plants used by tribal communities in Rajasthan, India. Journal of Ethnobotany and Traditional Knowledge. 2013;**9**(2): 115-122

[63] Tuso P, Stoll SR, Li WW. A plant-based diet, atherogenesis, and coronary artery disease prevention. The Permanente Journal. 2015;**8**(19):62

[64] Medak B, Singha LB. Trace elements and antioxidant activity of six wild edible plants that are widely consumed by ethnic tribes of Arunachal Pradesh, India. Indian Journal of Agricultural Research. 2018;**52**:85-88

[65] Pegu R, Gogoi J, Tamuli AK, Teron R. Ethnobotanical study of wild edible plants in Poba reserved Forest, Assam, India: Multiple functions and implications for conservation. Research Journal of Agriculture and Forestry Sciences. 2013;**3**:1-10

[66] Keskin C. Medicinal plants and their traditional uses. Plant Biology. 2018;**1**: 8-12

[67] Mahapatra AK, Mishra S, Basak UC, Panda PC. Nutrient analysis of some selected wild edible fruits of deciduous forests of India: An explorative study towards non-conventional bio-nutrition. Advance Journal of Food Science and Technology. 2012;**4**:15-21

[68] Worthington V. Nutritional quality of organic versus conventional fruits, vegetables, and grains. Journal of Alternative and Complementary Medicine. 2001;**7**:161-173

[69] Seal T. Determination of nutritive value, mineral contents and antioxidant activity of some wild edible plants from Meghalaya state, India. Asian Journal of Applied Sciences. 2011;**4**:238-246

[70] Holmelin NB. National specialization policy versus farmers' priorities: Balancing subsistence farming and cash cropping in Nepal. Journal of Rural Studies. 2021;**83**:71-80

[71] Rajkumari R, Singh PK, Das AK, Dutta BK. Ethnobotanical investigation of wild edible and medicinal plants used by the Chiru tribe of Manipur, India. Pleione. 2013;**7**:167-174

[72] Bultosa G, Molapisi M, Tselaesele N, Kobue-Lekalake R, Haki GD, Makhabu S, et al. Plant-based traditional foods and beverages of Ramotswa Village, Botswana. Journal of Ethnic Foods. 2020; **7**:1-15

Chapter 10

# The Characteristics of Popular Hardwood Species in Sudan-Review

*Fath Alrhman A.A. Younis and Emad H.E. Yasin*

## Abstract

Sudan's diverse ecosystems support valuable hardwood species crucial for economic, environmental, and social sustainability. This review evaluates seven key species such as *Acacia nilotica*, *Anogeissus leiocarpus*, *Azadirachta indica*, *Balanites aegyptiaca*, *Sclerocarya birrea*, *Khaya senegalensis*, and *Eucalyptus camaldulensis*, emphasizing their physical, mechanical, anatomical, and chemical properties. Findings indicate *A. leiocarpus* possesses the highest density (0.94 g/cm$^3$) and superior strength, ideal for construction, while *S. birrea* offers better workability despite lower density. These hardwoods support timber, fuelwood (16.7 million m$^3$ annually), and medicinal uses, yet they face severe threats from deforestation, agricultural encroachment, and climate change. Overexploitation risks long-term sustainability, necessitating urgent conservation measures. By synthesizing published data, this study highlights species-specific applications, from structural uses to pulp production and underscores the need for science-based forest management. The review provides critical insights for policymakers to balance utilization with preservation, ensuring Sudan's hardwood resources remain viable for future generations while supporting livelihoods and ecological stability. Sustainable practices must integrate research findings to mitigate current pressures on these indispensable species.

**Keywords:** Sudanese hardwood species, wood properties and utilization, forest resource management, sustainable timber production, deforestation and conservation

## 1. Introduction

Sudan's vast and ecologically diverse landscapes harbor a rich variety of hardwood species that constitute a vital natural resource for the nation. Spanning approximately 1.86 million square kilometers, Sudan's unique geographical position between the Sahara Desert and tropical savannas creates distinct ecological zones supporting diverse forest ecosystems [1]. These forests, comprising over 3156 species from 170 plant families [2], play a crucial role in sustaining livelihoods and maintaining ecological balance. Among these, seven key hardwoods—*Acacia nilotica*, *Anogeissus leiocarpus*, *Azadirachta indica, Balanites aegyptiaca, Sclerocarya birrea, Khaya senegalensis*, and *Eucalyptus camaldulensis*—have emerged as particularly significant due to their multiple uses and ecological importance [3].

The economic and ecological significance of these species is profound. They serve as primary sources of timber, fuelwood, and various non-timber forest products, with wood fuel alone accounting for 94% of Sudan's total forest production in 2022 [4]. These hardwoods provide essential materials for construction, furniture manufacturing, and traditional medicine while also functioning as carbon sinks and soil stabilizers against desertification [5]. Their multipurpose nature makes them indispensable to rural communities and national industries alike, supporting everything from household energy needs to pharmaceutical applications [6].

Despite their importance, Sudan's hardwood resources face severe and growing threats. Rapid population growth, agricultural expansion, and unsustainable harvesting practices have led to alarming deforestation rates [4]. Climate change exacerbates these pressures through altered rainfall patterns and rising temperatures, affecting species distribution and growth [7]. Compounding these challenges is a critical lack of comprehensive scientific data on the properties and sustainable management of these species. Current research remains fragmented, with studies often limited to specific regions or focusing on isolated characteristics [8, 9]. This knowledge gap hinders the development of effective conservation strategies and optimal utilization practices.

The justification for this study lies in addressing these critical gaps. A systematic review of hardwood properties is urgently needed to inform sustainable management decisions and maximize their economic potential. Understanding variations in density, mechanical strength, and durability can guide appropriate industrial applications [10], while documenting chemical composition can reveal new pharmaceutical and industrial uses [3]. Furthermore, synthesizing available ecological data can enhance conservation efforts amidst climate change [11]. Without such comprehensive knowledge, Sudan risks continued overexploitation and irreversible loss of these vital resources.

This chapter aims to provide a thorough review of the physical, mechanical, anatomical, and chemical properties of Sudan's seven most important hardwood species. The objectives are threefold: (1) to systematically compile and analyze existing research on these species, identifying patterns and inconsistencies; (2) to evaluate their suitability for various applications based on documented properties; and (3) to highlight critical knowledge gaps requiring further research. By achieving these objectives, this work will serve as a valuable resource for policymakers, industries, and conservationists working to balance utilization with preservation of Sudan's precious hardwood resources.

## 2. Methodology

The methodology for this review involved extensive literature searches across multiple academic databases including Google Scholar, ResearchGate, and ScienceDirect [7]. Studies published between 1987 and 2023 were considered, with particular attention to peer-reviewed journal articles, theses, and technical reports. Data on wood properties were extracted, compared, and analyzed to identify trends and variations across species and geographical locations. Where multiple studies reported conflicting results, average values were calculated to provide representative estimates [12].

## 3. Result and discussion

### 3.1 Present status of wood products in Sudan

In 1994, Sudan's total firewood consumption was estimated at approximately 5.9 $Mm^3$. A significant portion of this about 73.1% was consumed in low rainfall zones, reflecting the high dependence on wood fuel in ecologically vulnerable areas. The per capita firewood consumption was 0.36 $m^3$, with the desert region exhibiting the highest per capita use at 0.43 $m^3$. Moreover, low-income households accounted for 42.7% of total firewood consumption, underscoring the socioeconomic reliance on forest resources for domestic energy. Between 1999 and 2001, Sudan's annual industrial roundwood production averaged around 2.17 $Mm^3$, while fuelwood production rose to approximately 16.7 $Mm^3$. Charcoal demand remained consistently high during this period, largely dependent on stem wood—a practice known for substantial conversion losses [13]. By 2022, forest production in Sudan had undergone significant changes. Industrial roundwood production declined sharply to 163 $m^3$. However, the country produced 383,000 $m^3$ of wood chips and particles, 301,000 $m^3$ of pulpwood (round and split), and around 360,000 $m^3$ of sawlogs and veneer logs. Wood fuel remained the dominant forest product, with an estimated output of 15,583,393 $m^3$—representing approximately 94% of total forest production [4]. These figures are presented in **Figure 1**.

Hardwood species are trees that have broad leaves and produce seeds with some form of covering, such as fruits or nuts, which have needle-like leaves and produce seeds without any covering [14]. Hardwood has distinct characteristics such as high density, high durability, and high tensile and compressive strength, which makes it suitable for structural applications, such as in construction and furniture manufacturing. Also, it has attractive grain patterns, rich colors, and a polished appearance. This makes it highly desirable for use in high-end furniture, cabinetry, and decorative finishes. Moreover, hardwoods are generally easier to work with when it comes to cutting, shaping, and joining. They can be finely crafted, making them a popular choice for woodworking and furniture-making projects. In addition to their versatility, hardwoods are also known for their resilience in outdoor environments and their natural resistance to pests and decay. The examination of wood characteristics is

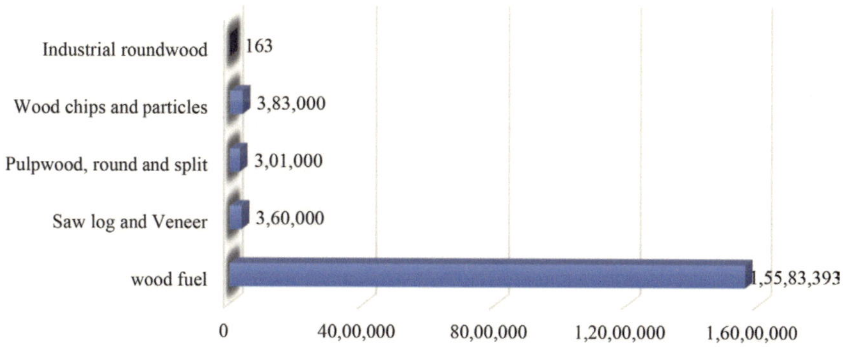


**Figure 1.**
*The forest wood production value ($m^3$) in 2022.*

a crucial aspect in assessing the appropriateness and utilization of wood material, which is dependent on the specific wood species [15]. To understand the behavior of wood, it is essential to conduct experiments that assess its physical and mechanical properties, in addition to examining its anatomical structure and chemical composition, both of which significantly influence its overall quality.

Hardwood species are important for Sudan for several reasons: first they provide valuable products such as timber, fuelwood, charcoal, fruits, nuts, honey, gum Arabic, and medicinal plants. These products support the livelihoods of many rural communities and contribute to the national economy. Secondly, they protect the land from desertification, erosion, and climate change by stabilizing the soil, retaining moisture, moderating temperatures, and enhancing biodiversity. Finally, they offer environmental services such as carbon sequestration, water purification, flood control, and wildlife habitat. However, hardwood species in Sudan are facing many threats, such as deforestation, overexploitation, agricultural expansion, fire, pests, diseases, and conflicts.

### 3.2 The most popular hardwoods used in Sudan are

- *Acacia nilotica subspecies nilotica (Sunt)* is a member of the Mimosoideae subfamily of leguminous trees, widely distributed across arid and semiarid regions worldwide. In Sudan, it grows along the banks of the Nile and in seasonal watercourses in areas like Blue Nile, Kordofan, Darfur, and Northern Bahr El Ghazal [2]. This medium-sized tree, reaching up to 17 meters in height and 50 cm in diameter, has a rounded or umbelliform crown with a straight, short cylindrical bole. Its blackish-gray, rough bark flakes in fibrous scales. The sapwood is white, while the heartwood is deep red-brown, hard, heavy, and durable. A fast-growing, drought-resistant legume, it plays a key role in biological nitrogen fixation, with a strong taproot system [16] and an extended growing season. This tree is utilized for timber, firewood, charcoal, gum Arabic, tannin, medicine, and fodder [3], with its hard, durable wood used for fuel, railway sleepers, structural components, transmission poles, and agricultural tools [17].

- *Anogeissus leiocarpus (Sahab)*, a member of the Combretaceae family, can grow up to 20 meters tall. It is commonly found in South Kassala, Kordofan, South Darfur, and Blue Nile states [2]. The grayish-brown wood is hard and ring-porous and contains surface crystals and traumatic ducts [18]. It is used locally for poles, beams, and fence posts, as well as for firewood and charcoal due to its high energy content.

- *Azadirachta indica (Neem)*, a fast-growing tree native to Asia, can reach heights of 15–20 meters, occasionally growing up to 35–40 meters. It is evergreen, although it sheds leaves during severe drought. The tree has a well-developed root system and produces a round or oval crown. The bark is fissured and varies in color from whitish gray to reddish-brown. Its wood is reddish and durable, making it suitable for various uses, including timber, fuel, and medicinal applications.

- *Balanites aegyptiaca (Heglig)*, belonging to the Zygophyllaceae family, is a drought-resistant tree typically found in arid and semiarid regions. It can grow 8–10 meters tall and is widely distributed in Sudan. The wood is hard, compact, and resistant to insects, making it suitable for agricultural implements, furniture, and tool handles. It is also valued for its medicinal properties, with various extracts used for antioxidant, antifungal, and anticancer treatments [19, 20].

- *Sclerocarya birrea (Humeid)*, a member of the *Anacardiaceae* family, grows up to 12 meters tall [19]. The wood is traditionally used for carving, furniture, and saddles. It is found in several regions of Sudan, including Kassala and Kordofan. The wood is soft, diffuse-porous, and medium-density, with a color ranging from grayish to reddish-brown. It is essential for sustaining rural livelihoods [21].

- *Khaya senegalensis (Mahogany)* is a medium-sized tree from the Meliaceae family, native to Africa. It can reach 15–30 meters in height with a diameter of up to 1 meter [22]. The bark is gray-brown, and the heartwood is pinkish-red. Its wood is often used in carpentry, furniture-making, and building construction, prized for its ease of use and good gluing properties [23].

- *Eucalyptus camaldulensis (Ban)*, an exotic species from Australia, can grow up to 30 meters tall. It is commonly planted in Sudan along riverbanks and irrigated plantations. The wood is used for timber, firewood, charcoal, pulp, paper, oil, and medicine [3]. Its white flowers are attractive to bees, and its trunk is straight, covered in rough, fibrous bark.

The review of Sudanese hardwood properties is vital for informing sustainable forest management, industrial applications, and environmental conservation. It is crucial for decision-making, ensuring the responsible and sustainable use of these valuable resources. This paper aims to provide a comprehensive overview of the properties of seven prominent hardwood species in Sudan, highlighting their economic and ecological importance.

### 3.3 Studies on tree species in Sudan: A focus on indigenous and exotic species

**Figure 2** indicates that *Acacia nilotica* is the most widely studied species across different regions in Sudan, which could be due to its widespread use in forestry, agriculture, and its role in agroforestry systems. Conversely, species like *Azadirachta indica* and *Khaya senegalensis* have been studied less, possibly due to their more specific or limited applications in Sudan's landscape. The distribution of studies also

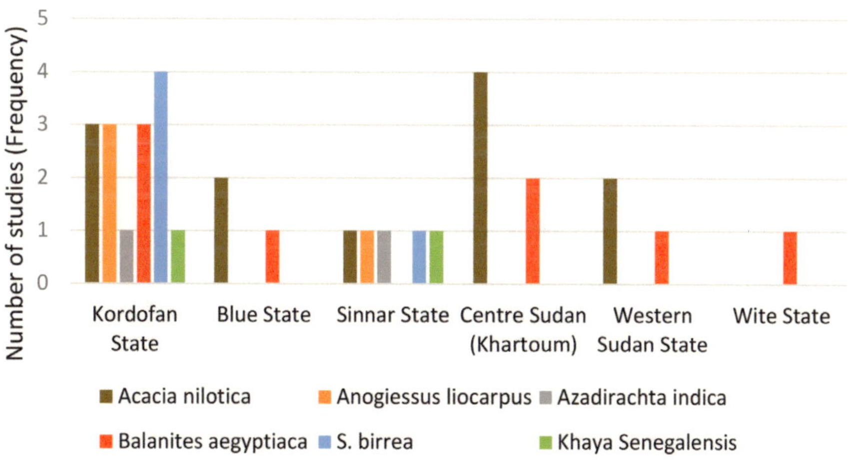


**Figure 2.**
*Geographical distribution of studies on tree species in Sudan.*

suggests regional preferences or ecological factors influencing the selection of species for research, with more studies concentrated in central and western parts of Sudan.

## 3.4 Comprehensive analysis of anatomical, physical, and mechanical properties across various tree species

The research studies reviewed in this paper provide valuable insights into the wood properties of several tree species, focusing on their anatomical, physical, and mechanical characteristics. These studies, summarized in **Tables 1–7**, contribute to

| Author and publication year | Study area | Title | Objective |
|---|---|---|---|
| Gamal et al. [9] | Kordofan (South) & Sinnar state | Suitability of some Acacia species grown in Sudan for paper making | The aim is to analyze the wood fiber properties of five Acacia species cultivated in Sudan and evaluate their potential for use in the pulp and paper industry. |
| Ahmed et al. [24] | Kordofan (North and South) | Effect of wood extractives on the equilibrium moisture content of six Sudanese hardwood species | To examine how the removal of hot-water extractives influences the desorption and adsorption equilibrium moisture content of six hardwood species. |
| Mohmmed et al. [11] | Darfur (North) | Relationship between anatomical properties and some physical and mechanical properties for five wood species growing in North Darfur-Sudan | To evaluate the relationships between anatomical properties and selected physical and mechanical characteristics of five wood species from North Darfur, Sudan. |
| Elamin [25] | Blue Nile | Mechanical properties (bending and Compression stress) as a tool for grading stress of *Acacai nilotica* wood. | To assess the mechanical properties of *Acacia nilotica* (L.) Willd and classify the grades based on basic stress. |
| Elgailani and Ishak [26] | Centre Sudan (Khartoum) | Determination of tannins of three common Acacia species of Sudan | To analyze and compare the tannin content of three common Acacia species in Sudan, as vegetable tannins play a significant role in the leather industry. |
| Hagar [27] | Centre Sudan (Khartoum) | Wood and Charcoal Anatomy of Eight Charcoal-producing Wood Species in Central Sudan | To examine the wood and charcoal anatomy of charcoal-producing tree species and explore the changes in wood macrostructure caused by charring. |
| Mahdi et al. [6] | Centre (Khartoum) & western Sudan | Characterization of *Acacia nilotica* as an indigenous tanning material of Sudan | To examine the quantity and quality of tannins in the bark and fruits of the three subspecies of *A. nilotica.* |
| Mahgoub, [28] | Kordufan (South) | Variation in wood density and shrinkage of twelve species from Kordofan state | To analyze the variation in wood density and shrinkage across twelve tree species from Kordofan state. |
| Khristova and Karar [29] | Centre Sudan (Khartoum) | Soda-anthraquinone pulp from three *Acacia nilotica* subspecies | To investigate the wood characteristics of the three Sudanese *Acacia nilotica* subspecies and their soda-AQ pulping potentialities. |
| Dafaalla and Ahmed [30] | Blue Nile | Variation in the mechanical properties of sunt wood (*Acacia Nilotica.*) from different sites growing in Blue Nile Sudan | To investigate the variation on mechanical properties of sunt wood (*Acacia nilotica* L. Wild-Del) from different sites. |

**Table 1.**
*List of research done on* A. nilotica *subspecies nilotica.*

| Author and publication year | Study area | Title | Objective |
|---|---|---|---|
| Abdelatif et al. [8] | Kordofan (South) | Structural characterization of three types of woods in Sudan: (*Anogeissus leiocarpus*) Sahab, (*Balanites aegyptiaca*) Heglieg, and (*Sclerocarya birrea*) Humeid | To determine structural characterization (i.e., strength classes) of some common Sudanese timber species in order to facilitate their use in structural design. |
| Younis et al. [31] | Kordofan (North) | Inter- and intraspecific differences in physical and mechanical properties of wood from *Sclerocarya birrea and Anogeissus leiocarpus* | To investigate the variability of wood along the height of a tree and horizontal positions (innerwood, middlewood, and outerwood). |
| Gamal et al. [12] | Kordofan (South)& Sinnar states | Variation in wood fiber characteristics among thirty-two hardwood species grown in low-rainfall woodland Savannah (Sudan) | To shed some lights on the variations in fiber characteristics of thirty-two hardwood species belonging to eighteen families grown in the low rainfall woodland savannah of Sudan |
| Osman [32] | | Variation in density and fiber dimensions and the demarcation of juvenile and nature wood of ten trees species grown in Sudan | To identify the transition point from juvenile to mature wood by examining radial variations in these properties. |

**Table 2.**
*List of research done on* A. leiocarpus.

| Author and publication year | Study area | Title | Objective |
|---|---|---|---|
| Gamal et al. [12] | Kordofan (South) & Sinnar states | Variation in wood fiber characteristics among thirty-two hardwood species grown in low-rainfall wood land Savannah (Sudan) | To shed some lights on the variations in fiber characteristics of thirty-two hardwood species belonging to eighteen families grown in the low rainfall woodland savannah of Sudan |

**Table 3.**
*List of research done on* A. Indica.

a better understanding of the variations within and between individual trees and the relationships between different wood properties. This information is crucial for both the practical utilization of the wood and the broader ecological and commercial implications of these species.

**Table 1** presents studies on *Acacia nilotica* subspecies *nilotica*, which extensively investigated the anatomical properties, chemical composition, extractive content, and both physical and mechanical properties of the wood. These studies also explored the variations in these properties and their relationships with anatomical features. This comprehensive approach provides essential insights into how the anatomical structure of the wood influences its physical and mechanical properties, which are important for its potential applications.

**Table 2** focuses on *Acacia leiocarpus*, where previous research investigated the physical and mechanical properties of the wood, along with their variations, as well

| Author and publication year | Study area | Title | Objective |
|---|---|---|---|
| Abdelatif et al. [8] | Kordofan (South) | Structural characterization of three types of woods in Sudan: (*Anogeissus leiocarpus*) Sahab, (*Balanites aegyptiaca*) Heglieg, and (*Sclerocarya birrea*) Humeid | To determine structural characterization (i.e., strength classes) of some common Sudanese timber species in order to facilitate their use in structural design. |
| Gamal et al. [33] | Kordofan (North & South) Blue state & White state | *Balanites aegyptiaca*: A multipurpose tree species for forest-based industry development in Sudan | To investigate some wood properties of *Balanites aegyptiaca* and to assess its suitability for pulp and paper and flooring industries |
| Mohammed et al. [11] | Darfur (North) | Relationship between anatomical properties and some physical and mechanical properties for five wood species growing in North Darfur-Sudan | To conducted to assess the relationships between anatomical properties and some physical and mechanical properties of five wood species growing in North Darfur, Sudan. |
| Awad [34] | Centre Sudan (Khartoum) | Variation among trees, sites, and log position along the stem on physical and mechanical properties of Heglig wood (*Balanites aegytica*) growing in central Sudan | To investigate the differences in the properties among population of *Balanites aegyptica* collected from Tozi, Elgazair, and Fraish forests |
| Hagar [27] | Centre Sudan (Khartoum) | Wood and charcoal anatomy of eight charcoal-producing wood species in Central Sudan | To study the wood and charcoal anatomy of charcoal-producing tree species and to investigate the alteration in wood macrostructure induced by charring. |
| Osman [32] | Blue Nile State, Northern Kordofan State, Southern Kordofan state and White Nile state | Variation in density and fiber dimensions and the demarcation of juvenile and mature wood of ten trees species grown in Sudan | To identify the transition point from juvenile to mature wood by examining radial variations in these properties. |
| Mahgoub [28] | Kordufan (South) | Variation in wood density and shrinkage of twelve species from Kordofan state | (1) To investigate the variation in wood density and shrinkage of the twelve hard wood species grows in Southern Kordofan, (2) study the relationship between shrinkage and moisture content as influenced by wood density, and (3) study the relationship between wood density, shrinkage, and anatomical characteristic. |

**Table 4.**
*List of research done on* B. aegyptiaca.

as its anatomical structure. These studies contribute to understanding the relationships between anatomical features and the physical and mechanical properties, offering useful data for potential industrial uses of *A. leiocarpus* wood. The variation in properties across different trees also indicates that the species may exhibit a range of wood qualities, which could be important for different applications.

| Author and publication year | Study area | Title | Objective |
|---|---|---|---|
| Abdelatif et al. [8] | Kordofan (South) | Structural characterization of three types of woods in Sudan: (*Anogeissus leiocarpus*) Sahab, (*Balanites aegyptiaca*) Heglieg, and (*Sclerocarya birrea*) Humeid | To determine structural characterization (i.e., strength classes) of some common Sudanese timber species in order to facilitate their use in structural design. |
| Younis et al. [31] | Kordofan (North) | Inter- and intraspecific differences in physical and mechanical properties of wood from *Sclerocarya birrea and Anogeissus leiocarpus* | To investigate the variability of wood along the height of a tree and horizontal positions (innerwood, middlewood, and outerwood). |
| Gamal et al. [12] | Kordofan (South) & Sinnar states | Variation in wood fiber characteristics among thirty-two hardwood species grown in low-rainfall woodland Savannah (Sudan) | To shed some lights on the variations in fiber characteristics of thirty-two hardwood species belonging to eighteen families grown in the low rainfall woodland savannah of Sudan. |
| Mahgoub [28] | South Kordufan | Variation in wood density and shrinkage of twelve species from Kordofan state | (1) To investigate the variation in wood density and shrinkage of the twelve hard wood species grows in Southern Kordofan, (2) study the relationship between shrinkage and moisture content as influenced by wood density, and (3) study the relationship between wood density, shrinkage, and anatomical characteristic. |
| Nasroun and Elzaki [10] | Sudan | The relationship between the anatomical structure and the mechanical properties of wood. | To examine the relationships between wood anatomical characteristics—quantified using stereological methods—and both the specific gravity and mechanical properties of wood in eight broad-leaved tree species cultivated in Sudan. |

**Table 5.**
*List of research done on* S. birrea.

| Author and publication year | Study area | Title | Objective |
|---|---|---|---|
| Gamal et al. [12] | Kordofan (South) & Sinnar states | Variation in wood fiber characteristics among thirty-two hardwood species grown in low-rainfall woodland Savannah (Sudan) | To shed some lights on the variations in fiber characteristics of thirty-two hardwood species belonging to eighteen families grown in the low rainfall woodland savannah of Sudan |

**Table 6.**
*List of research done on* K. senegalensis.

Research on *Acacia indica*, as shown in **Table 3**, primarily focused on the anatomical properties of the wood. Although the scope of this research was limited compared to the other species, the findings provide foundational knowledge of the wood's structure. Understanding these anatomical characteristics is a crucial first step toward evaluating its suitability for various uses, even though further studies would be needed to explore its physical and mechanical properties.

| Author and publication year | Study area | Title | Objective |
|---|---|---|---|
| Nasroun and Elzaki [10] | Sudan | The relationship between the anatomical structure and the mechanical properties of wood. | To examine the relationships between wood anatomical characteristics—quantified using stereological methods—and both the specific gravity and mechanical properties of wood in eight broadleaved tree species cultivated in Sudan. |

**Table 7.**
*list of research done on* E. camaldulensis.

**Table 4** highlights research on *Balanites aegyptiaca*, where both physical and mechanical properties, as well as anatomical features, were examined. Additionally, the variations in these properties both within and between trees were explored. This research helps to establish a more detailed understanding of how these properties differ among trees, which is critical for determining the species' practical applications. The study of such variations may also assist in developing better management strategies for the species.

In **Table 5**, studies on *Sclerocarya birrea* focused on the basic physical, mechanical, and anatomical properties of the wood while also exploring the variations in these properties across individual trees. This research offers valuable insights into the wood's overall quality and potential uses, providing a comprehensive view of how these properties fluctuate within the species. Understanding these variations can help optimize the use of *S. birrea* wood in various industries.

Research on *Kombo senegalensis*, presented in **Table 6**, was focused exclusively on the anatomical properties of the wood. While limited in scope, this study provides important baseline data that can serve as a foundation for future research, particularly for those interested in examining the species' physical and mechanical properties in more detail.

Finally, **Table 7** covers studies on *Eucalyptus camaldulensis*, where both anatomical and mechanical properties were examined, along with their interrelationships. The research highlights the importance of understanding how anatomical structure influences mechanical performance, which is crucial for determining the species' suitability for various industrial applications.

These findings are critical for assessing the practical uses of these woods and for developing sustainable management practices. Further research, particularly on the variations in wood properties within individual trees and between different regions, would enhance the knowledge base and improve the application of these species in various sectors.

### *3.4.1 The physical properties*

The physical properties of the seven Sudanese hardwood species, as presented in **Table 8**, reveal notable variations in the basic density, shrinkage values, and other wood characteristics, which are crucial for understanding the suitability of these species for different uses. Among the species studied, *Acacia leiocarpus* exhibited the highest basic density, followed by *Acacia nilotica* subspecies *nilotica*. On the other hand, *Sclerocarya birrea* showed the lowest basic density value, which indicates it may be less dense and potentially more lightweight compared to the other species, impacting its mechanical strength and durability.

DOI: http://dx.doi.org/10.5772/intechopen.1010985

| Physical property | Hardwood species | | | | | | |
|---|---|---|---|---|---|---|---|
| | *A. nilotica subs. Nilotica* | *A. leiocarpus* | *A. indica* | *B. aegyptiaca* | *S. birrea* | *K. senegalensis* | *E. camaldulensis* |
| Basic density g/cm$^3$ | 0.623–0.95 (0.79) | 0.92–0.96 (0.94) | | 0.66–0.78 (0.72) | 0.49–0.54 (0.52) | | 0.58 |
| Air dry density g/cm$^3$ | | | | 0.789 | | | |
| Oven-dry density | | | | | 0.55 | | |
| Green density | | | | | 1.09 | | |
| Bark to wood mass (%) | 14 | | | | | | |
| Bark to wood volume (%) | 12 | | | | | | |
| Tangential shrinkage % | 11.62 | | | 11.53 | | | 12.15 |
| Longitudinal shrinkage % | 0.86 | | | 1.59 | | | 1.19 |
| Radial shrinkage % | 5.28 | | | 5.62 | | | 6.26 |

*() = average value.*

**Table 8.**
*The physical properties drown from the literature.*

The research highlighted that oven-dry density and green density values were only available for *S. birrea*, while air-dry density values were reported for *Balanites aegyptiaca*. The lack of data for some species suggests that more comprehensive studies are needed to fill these gaps and provide a complete picture of the physical properties of these hardwoods. These density values are essential in determining how the wood behaves in terms of strength, weight, and its suitability for specific applications such as construction, furniture making, or fuel. Shrinkage characteristics are also important for understanding the dimensional stability of wood as it dries. *S. birrea* showed a tangential shrinkage of 11.53%, while *Balanites aegyptiaca* exhibited a slightly higher value of 12.15%. The radial shrinkage values were also reported for *S. birrea* (5.28%) and *B. aegyptiaca* (5.62%), with both species demonstrating relatively moderate shrinkage. *Acacia nilotica* subspecies *nilotica* exhibited a tangential shrinkage of 11.62%, similar to *S. birrea*, while *K. senegalensis* and *Eucalyptus camaldulensis* demonstrated somewhat higher shrinkage percentages in the radial direction. These shrinkage values are important in assessing the wood's potential for warping or cracking as it dries, affecting its suitability for specific end uses. Furthermore, the bark-to-wood mass and volume ratios were also considered, with *Acacia nilotica* subspecies *nilotica* showing a bark-to-wood mass percentage of 14%, which is relatively high compared to other species. These ratios indicate the proportion of bark relative to the usable wood and are an important consideration for wood processing and biomass utilization (**Table 8**).

Overall, the physical properties presented for these species underscore the diversity in wood characteristics, which can influence their applications in different industries. While species like *Acacia leiocarpus* and *Acacia nilotica* subspecies *nilotica* are denser and potentially stronger, species like *Sclerocarya birrea* and *Balanites aegyptiaca* may offer advantages in lightweight applications or where lower density is desired. Further research is needed to better understand the full spectrum of physical properties across these species, including the missing data for certain species, to guide their effective use in various industries.

### *3.4.2 The anatomical properties*

The anatomical properties of the seven Sudanese hardwood species, as presented in **Table 9**, provide essential information about their fiber and vessel characteristics,

| **Anatomical properties** | **Species** | | | | | | |
|---|---|---|---|---|---|---|---|
| ***Fiber*** | ***A. nilotica subs. Nilotica*** | ***A. leiocarpus*** | ***A. indica*** | ***B. aegyptiaca*** | ***S. birrea*** | ***K. senegalensis*** | ***E. camaldulensis*** |
| Fiber length (mm) | 1.0–2.10 (1.55) | 1.007–1.01 (1.01) | 0.83 | 1.15–1.77 (1.47) | 0.97–1.033 (1.00) | 1.19 | 0.81–2.28 (1.55) |
| Fiber width (μm) | 16.75–18.6 (17.86) | 12.4–14.2 (13.30) | 19.5 | 12.3–14.17 (13.24) | 15–28.4 (21.70) | 19.3 | 10–14.50 (12.25) |
| Lumen width (μm) | 7.23–7.80 (7.52) | 4.78–10.5 (7.64) | 10.27 | 6.87–10.5 (8.69) | 20.01 | 9.85 | 6.17 |
| Wall thickness (μm) | 5.3 | 4.69 | 4.61 | | 4.21 | 4.74 | |
| Fiber double wall thickness (μm) | 8.95–11.39 (10.17) | 1.9 | | 1.9–8.10 (5.00) | 2 | | 5–8.0 (6.50) |
| *Vessels* | 0.103 | | | | | | |
| Horizontal diameter (mm) | 0.120 | | | 0.083 | | | |
| Vertical diameter (mm) | 0.03 | | | 0.068 | | | |
| Horizontal double-cell wall (mm) | 0.02 | | | 0.027 | | | |
| Vertical double-cell wall (mm) | 0.10 | | | 0.012 | | | |
| Lumen diameter (mm) | 0.21 | | | 0.056 | | | |
| Mean horizontal free path (mm) | 0.23 | | | 0.693 | | | |
| Mean vertical free path (mm) | 28.54 | | | 0.844 | | | |
| Volume fraction % | 19 | | | 22 | | | |

*() = average value.*

**Table 9.**
*The anatomical properties drown from the literature.*

which directly influence the mechanical properties, processing efficiency, and potential applications of the wood. Variations in fiber length, fiber width, lumen width, and other anatomical features are important for determining the strength, durability, and suitability of the wood for specific uses, such as in construction, pulp production, or furniture making.

In terms of fiber length, *Acacia nilotica* subspecies *nilotica* exhibits the longest fiber, with a range of 1.0–2.10 mm and an average of 1.55 mm. This is relatively longer compared to *Acacia leiocarpus*, which has an average fiber length of 1.01 mm, and *Sclerocarya birrea*, which has fibers ranging from 0.97 to 1.033 mm. Longer fibers are generally associated with stronger and more durable wood, making *A. nilotica* subspecies *nilotica* potentially more suitable for applications requiring strength and durability. On the other hand, shorter fibers, such as those found in *A. indica* (0.83 mm), may be less mechanically robust but could be advantageous in applications where flexibility or lighter weight is desired.

Fiber width, another critical anatomical property, also shows variation across species. *A. nilotica* subspecies *nilotica* has fiber widths ranging from 16.75 to 18.6 μm, with an average of 17.86 μm. In comparison, *Balanites aegyptiaca* has a wider fiber width, with values ranging from 15 to 28.4 μm (average of 21.70 μm). Wider fibers typically contribute to higher wood density and improved mechanical strength, which may make *B. aegyptiaca* particularly useful in structural applications. *A. indica*, with a fiber width of 19.5 μm, also falls within the higher end of the range, suggesting it may have similar mechanical advantages.

Lumen width, which refers to the inner cavity of the fiber, is another important factor in determining the wood's porosity and processing characteristics. *Balanites aegyptiaca* has a particularly wide lumen diameter of 20.01 μm, much larger than the other species, such as *A. nilotica* subspecies *nilotica* (7.23–7.80 μm) or *S. birrea* (9.85 μm). A wider lumen can increase the wood's ability to absorb moisture and enhance its pulping potential, though it may reduce the wood's overall mechanical strength.

Wall thickness and double-wall thickness of fibers also vary significantly across species. For instance, the wall thickness of fibers in *A. nilotica* subspecies *nilotica* is 5.3 μm, while *A. leiocarpus* has a thinner fiber wall (4.69 μm). The thicker the fiber wall, the more robust and durable the wood, making species with thicker walls, such as *A. nilotica* subspecies *nilotica*, potentially more valuable for structural purposes. Similarly, the fiber double-wall thickness in *A. nilotica* subspecies *nilotica* ranges from 8.95 to 11.39 μm, significantly thicker than in *A. leiocarpus* (1.9 μm), which could contribute to differences in their overall wood strength and mechanical properties.

The vessel characteristics are also noteworthy. The horizontal and vertical diameters of vessels in *A. nilotica* subspecies *nilotica* are 0.120 mm and 0.03 mm, respectively, while in *Sclerocarya birrea*, the horizontal and vertical diameters are 0.083 mm and 0.068 mm, respectively. The size of the vessel elements, particularly the horizontal diameter, plays a role in the permeability and porosity of the wood, affecting its suitability for water conduction and use in applications where fluid movement is important. The larger horizontal diameter in *A. nilotica* subspecies *nilotica* may provide better conductivity compared to the smaller vessels in *S. birrea*.

Moreover, the volume fraction of fibers and vessels in the wood affects its overall composition. *A. nilotica* subspecies *nilotica* has a lower volume fraction of vessels (19%) compared to *Balanites aegyptiaca* (22%). This difference reflects the varying roles of fiber and vessel content in each species, with a higher volume fraction of fibers generally leading to denser and stronger wood. Conversely, species with a higher volume fraction of vessels may be more porous and lighter, which can influence their suitability for different commercial and industrial uses (**Table 10**). Overall, the anatomical properties of these seven Sudanese hardwood species exhibit considerable variability, which directly influences their mechanical properties and potential applications. Species like *Acacia nilotica* subspecies *nilotica* and *Balanites aegyptiaca* appear to have stronger, more robust fibers, making them suitable for construction and heavy-duty applications, while species like *Sclerocarya birrea* with smaller fiber diameters and vessel sizes may be better suited for lighter uses such as paper production or less demanding structural applications. Understanding these anatomical features helps to tailor the use of each species for specific purposes, whether in the timber industry, pulp and paper manufacturing, or other wood-based products. Further studies are needed to explore the relationships between these anatomical characteristics and the overall mechanical performance of the wood, providing a more detailed understanding of their full potential.

### *3.4.3 The mechanical properties*

**Table 10** provides a comprehensive overview of the mechanical properties of seven Sudanese hardwood species, which include important metrics such as Modulus of Rupture (MOR), Modulus of Elasticity (MOE), Compression Strength (CS), Shear Strength (SH), and other relevant mechanical characteristics. These properties are vital for understanding the suitability of each species for various engineering and construction applications, as well as for assessing their overall structural integrity and durability.

Modulus of Rupture (MOR), which is a measure of the wood's ability to resist bending or breaking under stress, varies significantly across the species. *Acacia nilotica* subspecies *nilotica* shows the highest MOR value at 140.70 MPa, indicating that this species has superior bending strength compared to the others. This suggests that *A. nilotica* subspecies *nilotica* would be highly suitable for structural applications where high bending strength is required, such as beams and supports. In comparison, *Balanites aegyptiaca* and *Sclerocarya birrea* have lower MOR values, 51.58 MPa and 99.03 MPa, respectively, making them less suitable for heavy-duty structural applications but potentially more suitable for lighter uses where lower strength is acceptable.

Modulus of Elasticity (MOE), which indicates the stiffness or rigidity of the wood under stress, shows a wide range of values. *A. nilotica* subspecies *nilotica* again leads with a value of 17,500.51 MPa, suggesting that its wood is particularly stiff and resistant to deformation. This is an important characteristic for wood used in construction, particularly in applications requiring rigidity and minimal deflection under load, such as in flooring or heavy structural elements. *A. leiocarpus* and *A. indica* also exhibit relatively high MOE values, indicating their suitability for similar applications. On the other hand, *B. aegyptiaca*, with an MOE value of 6932.7 MPa, demonstrates lower rigidity and may be more suitable for applications where some flexibility is beneficial, such as in crafts or composite materials.

Compression Strength (CS) parallel to MOE is another crucial factor for understanding the compressive load-bearing capacity of wood. The values for CS in the longitudinal direction (CS//) range from 38.26 MPa in *Balanites aegyptiaca* to 68.47 MPa in *Acacia nilotica* subspecies *nilotica*, with the latter exhibiting the highest compressive strength. This suggests that *A. nilotica* subspecies *nilotica* is ideal for load-bearing applications, particularly in structural and construction fields where the material will be subjected to vertical loads. In contrast, the lower CS values in species such as *B. aegyptiaca* indicate that they may be better suited for less load-intensive applications or for uses in which the material will not be heavily stressed in compression.

| Mechanical properties | Species | | | | | | |
|---|---|---|---|---|---|---|---|
| | *A. nilotica subs. Nilotica* | *A. leiocarpus* | *A. indica* | *B. aegyptiaca* | *S. birrea* | *K. senegalensis* | *E. camaldulensis* |
| MOR | 140.70 | 70.97–96.7 (83.84) | | 87.25–114.5 (102.38) | 40.73–62.43 (51.58) | | 99.03 |
| MOE | 9400–25601.02 (17500.51) | 8400–12,577 (10488.50) | | 8400–10,493 (94446.50) | 6600–7265.4 (6932.70) | | 8503.4 |
| CS // | 58.7–68.47 (63.59) | 45.77–66.57 (56.17) | | 47.47–60.12 (53.80) | 38.26–39.51 (38.89) | | 57.00 |
| SH // | 32.7–34.7 (33.70) | 16.52–16.64 (16.58) | | 19.45 | 13.32–16.66 (14.99) | | |
| Tension strength | | | | 52.19 | 27.85 | | |
| US | 234.64 | | | | | | |
| Ultimate CS | 102.62 | | | | | | |
| MOC | 2166.3–3446.63 (2806.47) | | | 1959.7 | | | 1727.5 |
| Poison ratio | 0.297 | | | | | | |
| H (cross surface) | | | | 87 | | | |
| H (radial surface) | | | | 46 | | | |

*() = average value of the range; UBS = Ultimate strength; H = Hardness; modulus of elasticity from the compression.*

**Table 10.**
*The mechanical properties drown from the literature.*

Shear Strength (SH), which measures the wood's resistance to forces that can cause sliding along the grain, is an essential property for assessing the wood's stability and potential for deformation under shear stress. The values of shear strength range from 13.32 MPa to 34.7 MPa, with *A. nilotica* subspecies *nilotica* and *A. leiocarpus* having higher values compared to other species. This indicates that these species may perform better in situations where shear forces are a concern, such as in framing or joinery where wood components are subjected to lateral loads.

The tensile strength, although available for only a few species, shows *A. indica* at 52.19 MPa and *S. birrea* at 27.85 MPa. These values are indicative of the species' ability to withstand pulling or stretching forces. A higher tensile strength suggests that *A. indica* might be more appropriate for applications where the wood is subject to tensile stresses, such as in rope or cable products, whereas the lower value for *S. birrea* suggests that it may be less suitable for such uses.

Ultimate compressive strength (US) and ultimate shear (UBS) are also important in evaluating the wood's performance under extreme stresses. The value for *A. nilotica* subspecies *nilotica* (234.64 MPa) suggests its high ultimate strength under compressive loads, reinforcing its suitability for applications where the wood is exposed to heavy loads. Similarly, the Hardness (H) of *A. indica* and *E. camaldulensis*, measuring 87 and 46 respectively, indicates their relative resistance to indentation, which is important in flooring and other high-contact applications.

Poisson's Ratio, which is a measure of the material's tendency to expand in directions perpendicular to the direction of compression, is reported as 0.297 for *A. nilotica* subspecies *nilotica*, which is within a typical range for hardwoods. This ratio reflects the expected deformation behavior of the wood under load and helps predict how the material will perform in practical applications (**Table 10**).

Overall, the mechanical properties of these Sudanese hardwood species show substantial variation, with *Acacia nilotica* subspecies *nilotica* demonstrating superior strength and stiffness across multiple metrics such as MOR, MOE, and CS. These properties make it an excellent candidate for structural and construction applications. Species like *Balanites aegyptiaca* and *Sclerocarya birrea*, with lower MOR and MOE values, may be better suited for lighter, less demanding applications. The variety in mechanical properties between these species suggests that each species has distinct advantages depending on the specific engineering requirements, such as strength, flexibility, and resistance to deformation. Future studies should explore how these mechanical properties correlate with the wood's long-term performance in real-world applications, such as in varying environmental conditions or under different load scenarios.

### *3.4.4 The chemical properties*

**Table 11** provides a detailed comparison of the chemical composition of seven Sudanese hardwood species. These species were analyzed for several key chemical properties, including ash content, silica content, tannins, extractives, cellulose, holocellulose, alpha-cellulose, pentosans, and lignin. These chemical constituents are important for understanding the wood's overall quality, durability, and potential uses in various industries, such as paper production, wood preservation, and bioenergy.

Ash Content is a key indicator of the mineral content in wood. Ash content is generally low across all species, with *Acacia nilotica* subspecies *nilotica* showing an ash content of 0.7%. This is typical for hardwood species, indicating a relatively low mineral content. High ash content can negatively affect the burning properties of wood and the efficiency of thermal treatments, making it less desirable for applications such as bioenergy production. The low ash content of these species suggests that they may be better suited for uses in which low mineral content is advantageous.

Silica content in wood is another important factor, particularly for its effect on the wood's processing, such as in the paper and pulp industries. *A. nilotica* subspecies *nilotica* shows a very low silica content of 0.1%, which suggests that this species may be less abrasive and easier to process compared to others with higher silica levels. Silica is often undesirable in papermaking as it can lead to wear and tear on machinery. As the silica content in these species is low, they may be suitable for industries that require low abrasiveness in raw materials.

Tannins are polyphenolic compounds that affect the wood's resistance to decay and its use in certain industries. The tannin content in *A. nilotica* subspecies *nilotica* is relatively high at 11.8%, which may contribute to its resistance to biological degradation. Tannins also play an important role in the leather and wood treatment industries due to their natural preservative qualities. High tannin content can make the wood less susceptible to fungal and insect damage, which could be beneficial for outdoor applications or in wood preservation treatments.

The extractives content refers to the chemical compounds that can be extracted from the wood using solvents like hot water, NaOH, and alcohol-benzene. *A. nilotica* subspecies *nilotica* has an extractive content of 11.8%, and *A. leiocarpus* shows a slightly higher range (11.03–17.83%, with an average of 14.43%). These extractives are important as they can influence the wood's performance in various applications.

| Chemical composition | Species | | | | | | |
|---|---|---|---|---|---|---|---|
| | *A. nilotica subs. Nilotica* | *A. leiocarpus* | *A. indica* | *B. aegyptiaca* | *S. birrea* | *K. senegalensis* | *E. camaldulensis* |
| Ash | 0.7 | | | | | | |
| Silica | 0.1 | | | | | | |
| Tannins | 11.80 | | | | | | |
| Extractives content | 11.8 | 11.03–17.83 (14.43) | | | 13.85 | | |
| Soluble in hot water | 5.7 | | | | | | |
| Soluble in NaOH | 16.5 | | | | | | |
| Soluble in alcohol-benzene | 5.5 | | | | | | |
| Cellulose, kurschner-Hoffer | 47.8 | | | | | | |
| holocellulose | 76.8 | | | | | | |
| Alpha-cellulose | 51.9 | | | | | | |
| Pentosans | 18.7 | | | | | | |
| Lignin | 26.8 | | | | | | |
| Cellulose/lignin | 1.8 | | | | | | |

**Table 11.**
*Chemical composition drowns from the literature.*

Extractives can affect the wood's durability, color, odor, and resistance to decay. Species with higher extractives may have better natural resistance to environmental stresses and could be more suitable for outdoor or marine applications where such properties are needed.

The solubility in solvents is a measure of how much of the wood's chemical compounds can be dissolved in specific solvents. *A. nilotica* subspecies *nilotica* shows solubility values of 5.7% in hot water, 16.5% in NaOH, and 5.5% in alcohol-benzene. These solubility measurements indicate the ease with which the wood's chemical components can be extracted and processed. High solubility in NaOH is particularly relevant for paper production, as alkali treatments are commonly used to break down lignin and hemicelluloses in wood fibers.

Cellulose and Hemicelluloses are major components of the wood's structure and are crucial for its strength and utility. The cellulose content in *A. nilotica* subspecies *nilotica* is reported as 47.8%, and the holocellulose content is 76.8%. Holocellulose refers to the combined content of cellulose and hemicelluloses. *A. nilotica* subspecies *nilotica* also shows a relatively high alpha-cellulose content of 51.9%, which is the most crystalline and strongest form of cellulose. Higher alpha-cellulose content typically correlates with better quality pulp and paper production, as alpha-cellulose is more easily processed. The high cellulose and holocellulose contents of this species suggest that it could be well-suited for pulp and paper production or for bioenergy applications.

Lignin, which serves as the glue that binds cellulose fibers in the wood, is another important component of the chemical makeup. *A. nilotica* subspecies *nilotica* contains 26.8% lignin, which is a moderate value for hardwood species. Lignin content is significant because it influences the wood's rigidity and resistance to decay, but it also impacts the wood's digestibility for bioenergy and its suitability for certain types of processing. Lower lignin content typically leads to easier processing, such as in the case of pulping for paper or biofuel production, but too little lignin can reduce the wood's strength and resistance to environmental stresses.

The cellulose-to-lignin ratio is an important indicator of the wood's suitability for specific applications. In this case, *A. nilotica* subspecies *nilotica* has a cellulose-to-lignin ratio of 1.8, which suggests a favorable balance between strength and processability. A higher ratio is generally desirable for paper and bioenergy production, as it indicates that the wood contains more cellulose relative to lignin, making it easier to break down for these purposes.

Overall, *A. nilotica* subspecies *nilotica* appears to have a well-rounded chemical profile that makes it suitable for a variety of industrial applications, particularly in paper production and bioenergy. Its high cellulose, holocellulose, and alpha-cellulose content, combined with its relatively low lignin and silica content, suggest that it can be processed efficiently for these uses. Other species such as *A. leiocarpus* and *S. birrea* also show promising chemical profiles, but *A. nilotica* subspecies *nilotica* stands out for its balanced chemical composition, which enhances its versatility for different industrial applications. Future research into the specific interactions between these chemical components and the wood's mechanical properties could provide valuable insights into optimizing the use of these species in various industries.

## 4. Conclusions

Sudanese hardwood species are vital to both economic and environmental sustainability, offering essential resources for local communities and industries. This study has examined the diverse properties of key hardwood species, focusing on their structural strength, durability, and adaptability to varying climatic conditions, which makes them suitable for timber, fuelwood, furniture, and medicinal applications. However, challenges such as deforestation, land-use change, and climate variability threaten their long-term sustainability. The review emphasizes the need for enhanced forest management strategies, research on wood properties, and conservation efforts to ensure a balance between resource use and environmental preservation. Future actions should focus on promoting reforestation, implementing sustainable harvesting practices, and strengthening local knowledge about the ecological and economic benefits of these species. By adopting effective forest governance and conservation measures, Sudan can secure the long-term viability of its hardwood resources while supporting both economic growth and environmental resilience. Additionally, future research should address key areas like the variation in wood properties across growth stages, the impact of environmental stressors, and the potential for these species in industries such as pulp and paper production, bioenergy, and agroforestry. Standardizing testing methodologies and conducting life cycle analysis (LCA) will provide reliable data for better land management decisions and ensure the sustainable utilization of these valuable hardwood species.

## Acknowledgements

We are grateful to our colleagues and friends for their meaningful discussions, valuable advice, and encouragement throughout the writing process. We also extend our sincere appreciation to the editors and reviewers for their constructive feedback, and to the Institute of Geomatics and Civil Engineering, University of Sopron, for their academic support and collaboration.

## Conflict of interest

The authors declare no conflict of interest.

## Author details

Fath Alrhman A.A. Younis[1,2] and Emad H.E. Yasin[3,4]*

1 Faculty of Wood Engineering and Creative Industries, Institute of Wood Technology and Technical Sciences, University of Sopron, Sopron, Hungary

2 Faculty of Forest Sciences and Technology, Department of Wood Science and Technology, University of Gezira, Wad Madani, AL Gezira State, Sudan

3 Faculty of Forestry, Department of Forest Management, University of Khartoum, Khartoum North, Sudan

4 Faculty of Forestry, Institute of Geomatics and Civil Engineering, University of Sopron, Sopron, Hungary

*Address all correspondence to: emad.yasin823@gmail.com

## References

[1] Sikainga AA, Al-Shahi AS, Sabr M, el Din S, Jay L. Unit, Economist Intelligence and Collins, Robert O. "Sudan". Encyclopedia Britannica; 2023. Available from: https://www.britannica.com/place/Sudan [Accessed: October 30, 2023]

[2] El Amin HM. Trees and Shrubs of Sudan. Ithaca, New York: Ithaca Press and University of Michigan; 1990. p. 484

[3] Robinson J. Useful wild tree resources of southern Sudan: A review. Plant Genetic Resources. 2006;**4**(3):188-197

[4] FAO. Forest product consumption and production. Major consumers of forest products. Percentage of Global Consumption. 2023:2020

[5] Mohammed AA, Bauer H, Coals P, Meina O, Elsadig I, Ahmed R, et al. Abundance of larger mammals in Dinder National Park, Sudan. European Journal of Wildlife Research. 2023;**69**(3):46

[6] Mahdi H, Palmina K, Glavtch I. Characterization of *Acacia nilotica* as an indigenous tanning material of Sudan. Journal of Tropical Forest Science. 2006;**18**(3):181-187

[7] Musa FI, Sahoo UK, Eltahir ME, Magid TDA, Adlan OE, Abdelrhman HA, et al. Contribution of non-wood forest products for household income in rural area of Sudan–a review. Journal of Agriculture and Food Research. 2023;**14**:100801

[8] Abdelatif AO, Abdrahman RA, Abdalla AMA. Structural characterization of three types of woods in Sudan; (Anogeissus leiocarpus) Sahab, (Balanites aegyptiaca) Heglieg, and (Sclerocarya birrea) Humeid. In: Structures. Amsterdam, Netherlands: Elsevier; 2022. pp. 351-357

[9] Gamal HMS, Abdalla AMA, Abdelgadir AAY, Gibreel HH. Suitability of some acacia species grown in Sudan for paper making. Agriculture and Forestry Journal. 2021;5(1):58-67

[10] Nasroun TH, Elzaki OT. The relationship between the anatomical structure and the mechanical properties of wood. Sudan Silva. 1987;**6**(26):88-99

[11] Mohammed NE, Abdallah A, et al. Relationship between anatomical properties and some physical and mechanical properties for five wood species growing in North Darfur-Sudan [PhD thesis]. Sudan: Sudan University of Science and Technology; 2019

[12] Gamal HMS, Abdelgadir Y, Azim A, Bues C-T. Variation in wood fiber characteristics among thirty-two hardwood species grown in low-rainfall wood land Savannah (Sudan). In: The 5th Conference on Hardwood Research and Utilisation in 2012. Sopron, Hungary: University of Sopron Press; 2012. p. 104

[13] Ali GE. Studies on Consumption of Forest Products in the Sudan—Woodfuel Consumption in the Household Sector. Khartoum, Sudan: Energy Research Institute; 1994. p. 199

[14] Desch HE, Dinwoodie JM. Timber, Structure, Properties, Conversion, and Use. 7th ed. London: Macmillan Press Ltd.; 1996. p. 297

[15] Koch P. Utilization of Hardwoods Growing on Southern Pine Sites. Vol. 605. no. 3. US Department of Agriculture, Forest Service; 1985. p. 3710

[16] Toky OP, Bisht P. Observations on the rooting patterns of some agro forestry trees in an arid region of north-western India. Agroforestry Systems. 1992;**18**:249-261

[17] Thirakul S. Manual of Dendrology. Quebec, Canada: Groupe Poulin, Theriault Ltee Consultants; 1984. pp. 245-266

[18] Jayeola AA, Aworinde DO, Folorunso AE. Use of wood characters in the identification of selected timber species in Nigeria. Notulae Botanicae Horti Agrobotanici Cluj-Napoca. 2009;**37**(2):28-32

[19] Vogt K. A Field Guide to the Identification, Propagation and Uses of Common Trees and Shrubs of Dryland Sudan. UK: SOS Sahel International; 1995

[20] Broun AF, Massey RE. Flora of the Sudan. The Controller, Sudan Govt. Office, Wellington House, Buchingham Gate, London, S.W.I. 1929

[21] Sahni KC. Important Trees of the Northern Sudan. United Nations and FAO; 1968

[22] Anon. Sokoto State. 2014. Available from: http://en.m.wikipedia.org/wiki/Sokoto_State [Accessed: June 05, 2025]

[23] Prota. Timbers/Bois d'œuvre 1: Khaya senegalensis (Desr.) A. Juss. Prota. 7(1). Available from: http://database.prota.org/PROTAhtml/khaya%20senegalensis_En.htm [Accessed: October 22, 2014]. (PDF) Variation in Density and Shrinkage between Sawmill and Hand processed khaya senegalensis wood in sokoto, North western Nigeria. 2014. Available from: https://www.researchgate.net/publication/277132271_Variation_in_Density_and_Shrinkage_between_Sawmill_and_Hand_processed_khaya_senegalensis_wood_in_sokoto_North_western_Nigeria [Accessed: June 05, 2025]

[24] Ahmed ZAA, Abdelgadir AY, Abdalla AMA. Effect of wood extractives on the equilibrium moisture content of six Sudanese hardwood species. University of Khartoum Journal of Agricultural Sciences. 2019;**25**(1):107-125

[25] Elamin EE. Mechanical properties (bending and compression stress) as a tool for grading stress of *Acacai nilotica* wood. Mechanical Properties (Bending and Compression Stress) as a Tool for Grading Stress of *Acacai Nilotica* Wood. 2018;**18.1**:8-8

[26] Elgailani IEH, Ishak CY. Determination of tannins of three common acacia species of Sudan. Advances in Chemistry. 2014;**2014**(1):1-5

[27] Hagar HMM. Wood and charcoal anatomy of eight charcoal-producing wood species in central Sudan [doctoral dissertation]. Sudan: University of Khartoum; 2010

[28] Mahgoub AAM. Variation in wood density and shrinkage of twelve species from Kordofan State [doctoral dissertation]. University of Khartoum; 2001

[29] Khristova P, Karar I. Soda-anthraquinone pulp from three *Acacia nilotica* subspecies. Bioresource Technology. 1999;**68**(3):209-213

[30] Dafaalla ME, Ahmed. Variation in the mechanical properties of Sunt Wood (*Acacia Nilotica*.) from different sites growing in Blue Nile Sudan [M.Sc. thesis]. Sudan: University of Khartoum; 1998

[31] Younis FA, Abdelgadir AY, Ahmed ZA, Govina JK, Németh R. Inter- and intraspecific differences in physical

and mechanical properties of wood from Sclerocarya birrea and Anogeissus leiocarpus. Acta Silvatica Et Lignaria Hungarica: An International Journal in Forest, Wood and Environmental Sciences. 2022;**18**(1):57-69

[32] Osman MEB. Variation in density and fiber dimensions and the demarcation of juvenile and nature wood of ten trees species grown in Sudan [doctoral dissertation]. Sudan: University of Khartoum; 2001

[33] Gamal HMS, Abualgasim MR, Gibreel HH, Abdalla AMA, Mohammed KFEE. *Balanites Aegyptiaca*: A Multipurpose Tree Species for Forest Based Industry Development in Sudan. Adv in Agri, Horti and Ento: AAHE-134; 2020

[34] Awad ME. Variation among trees, sites and log position along the stem on physical and mechanical properties of Heglig Wood (*Balanites Aegytica*) growing in Central Sudan [M.Sc. thesis]. Sudan: University of Khartoum; 2015